VISUAL POLLUTION

VISUAL POLLUTION

Concepts, Practices and Management Framework

BY

RAHEEL NAWAZ
Staffordshire University, UK

And

KHYDIJA WAKIL
National University of Sciences and Technology, Pakistan

emerald
PUBLISHING

United Kingdom – North America – Japan – India
Malaysia – China

Emerald Publishing Limited
Howard House, Wagon Lane, Bingley BD16 1WA, UK

First edition 2022

British Library Cataloguing in Publication Data
A catalogue record for this book is available from the British Library

ISBN: 978-1-80382-042-2 (Print)
ISBN: 978-1-80382-041-5 (Online)
ISBN: 978-1-80382-043-9 (Epub)

Printed and bound by CPI Group (UK) Ltd, Croydon, CR0 4YY

ISOQAR certified Management System, awarded to Emerald for adherence to Environmental standard ISO 14001:2004.

Certificate Number 1985
ISO 14001

INVESTOR IN PEOPLE

To Daniyal, life wouldn't mean much without you!

(Professor Raheel Nawaz)

To late Professor Ghulam Abbas Anjum, my first research mentor who encouraged me to pursue visual pollution research

(Dr Khydija Wakil)

QUOTATION

We have eyes, and we're looking at stuff all the time, all day long, and I just think that whatever our eyes touch should be beautiful, tasteful, appealing, and important. We need to emphasise the responsibility that designers and illustrators have towards the people they create things for; Whether it's a coffee cup, or a poster, or a book illustration, or a typeface, it has to be designed in such a way that it is not trashy, and doesn't pollute your eyes. We have so much pollution out in the air. Our eyes are being polluted. We have visual pollution out there, and I have a very strong sense about that.

Eric Carl (2014)

CONTENTS

LIST OF FIGURES

Chapter 4

Chapter 5

LIST OF BOXES

LIST OF TABLES

LIST OF ACRONYMS

AHP	Analytical Hierarchy Process
AI	Artificial Intelligence
CDG	City District Government
DHA	Defence Housing Authority
GDP	Gross Domestic Product
GIS	Geographic Information System
HDI	Human Development Index
IT	Information Technology
KML	Keyhole Markup Language
ODK	Open Data Kit
OSM	Open Street Maps
QSPM	Quantitative Strategic Planning Matrix
RDA	Rawalpindi Development Authority
SWOT	Strengths, Weaknesses, Opportunities, Threats
TMA	Tehsil Municipal Administration
UK	United Kingdom
USA	United States of America
VIA	Visual Impact Assessment
VP	Visual Pollution
VPA	Visual Pollution Assessment
VPI	Visual Pollution Index
VPO	Visual Pollution Object

ABOUT THE AUTHORS

Professor Raheel Nawaz PFHEA, is currently serving as the Pro Vice-Chancellor (Digital Transformation) at Staffordshire University, where he provides strategic leadership and management of the University's continuing digital transformation agenda and leads transformative action across the University to deliver enhanced academic experience. Prior to that he held two substantive roles at Manchester Metropolitan University. As the Director of Business Transformations Research Centre, he led nearly 200 researchers specialising in digital transformation, analytics, industry 4.0, transforming markets and transforming places. As the Director of Digital Technology Solutions, he led Met's award-winning cross-faculty Digital Degree Apprenticeships portfolio delivering UG/PG programmes to nearly 900 apprentices from 70-plus leading technology employers. He was also the founding Head of the Apprenticeships Research Unit and the Text & Data Mining Lab at Manchester Met. Professor Nawaz is a renowned expert in industry-academia co-creation, especially for high-impact work integrated programmes like Degree Apprenticeships. He has advised on the establishment and launch of such programmes in Spain, New Zealand, Canada and Pakistan. He has addressed the Westminster Education Policy Forum twice and has recently been commissioned by the QAA to systematically analyse the Degree Apprenticeships pedagogies across English universities and advise the sector on best academic practices. Professor Nawaz is also a leading researcher in Artificial Intelligence and Digital Education. He holds several adjunct professorships and scientific directorships across Asia and North America. He sits on the boards of research and charitable organisations like the National Centre for Artificial Intelligence (Pakistan), TechSkills (UK) and NTF (UK), and has advised national policy organisations including the Prime Minister's Task Force on Science and Technology (Pakistan). He has authored over 150 peer-reviewed research articles and his career grant capture stands at over £14 million. He has graduated 19 PhD students so far. According to Google Scholar, he is among the top-10 most cited scholars in the world in the fields of Digital Transformations, Applied Artificial Intelligence and Educational Data

Science. In the past, he has held senior leadership positions in the private further and higher education sector and was an army officer before that.

Dr Khydija Wakil is a professional Urban Planner and a Researcher in the fields of built environment, urban design and visual quality of the urban neighbourhoods and areas. She has a PhD in Urban Visual Pollution from the National University of Sciences and Technology, Pakistan. She is currently working as the Chief Executive Officer at City Pulse Pvt. Ltd, an urban planning consulting firm in Pakistan. She is an emerging international expert on visual pollution, and her research on conceptualisation of visual pollution and related assessment mechanisms has been published in leading journals and conferences. She has written several public engagement pieces on the topic as well and is a leading activist and ambassador for managing visual pollution.

ACKNOWLEDGEMENTS

We would like to thank the following colleagues and friends for their valuable input in refining the manuscript.

Dr Muhammad Qadeer ul Hussnain
Mr Paul Thompson
Dr Abdul Waheed
Mr Inam Ul Haq
Dr Jamal Uddin Thaheem
Dr Muhammad Fayyaz
Dr Afia Zubair Raja

1

VISUAL POLLUTION: A NEW ADDITION TO THE POLLUTION FORMS

1. INTRODUCTION

In published English literature, the first appearance of the term 'visual pollution' can be traced back to the early twentieth century, when it was introduced to highlight the secondary impacts of man-made interventions. In the late 1960s, the United States declared visual pollution as one of the most significant environmental problems of the era. Almost 60 years ago, Maclurcan published 'visual pollution' which focuses on the importance of visual aesthetics in ensuring urbanity and sanity within the built environments (Maclurcan, 1973). He illustrated his point of view by using the scenario of an ideal urban community with serene visual surroundings which subsequently loses its visual appeal due to the ugliness of visual pollution. He highlighted the exponentially changing identity of urban spaces and the plethora of issues resulting from dramatic trends towards urbanisation. Maclurcan characterised his fears about difficulties in handling visual elements in the environment as one of the most frightening prospects of future urban lives and hypothesised that visual pollution could have severe psychological implications on the population. He went on to argue that social and psychological problems such as trauma, depression, unfriendliness, suspicion and anxiety were increasing due to the spread of ugliness in the man-made environment. Maclurcan uses the lifecycle of flowers as a metaphor for all organisms, including humans, animals, aquatic lives, trees, plants and birds. The duration of this lifecycle may be shortened or prolonged according to the quality of life within its ecosystems. In particular, the interactions of the population of a location are the mirror of its man-made environment. He proposed serious concerns about the decay of the urban visual environment, with which its inhabitants are complacently accustomed.

In addition, he reflected on limitations related to the realisation, acceptance, quantification and management of visual pollution. He warned that delays in controlling, managing and regularising visual pollution could result in a physical and psychological catastrophe. Although he could not forecast 'how' this disaster would potentially play out, 60 years later, we can see the footprints of his concerns.

Building on the aforementioned argument, this chapter documents the realities and of how visual pollution has affected urban spaces since the publication of Maclurcan's work. We argue that the ugliness of ungoverned visual pollution has become a monster, which has swallowed the aesthetics of the built environment, and investigate whether the miseries of urban dwellers have increased due to the unchecked onslaught of visual pollution. Macluran's visual projections of disturbing skylines flooded by outdoor advertisements, monotonous building designs, destruction of the natural landscape, over-crowding of concrete jungles and littering etc. have proven to be accurate, and many of the problems, which first manifested themselves years ago in developed countries, are now equally affecting developing countries.

It is essential at this point to define the key concepts used within discussions of visual pollution, which include the *environment*, *aesthetics*, *front-stage* and *back-stage* of a city. The term 'environment' refers to the combination of physical objects, natural features, humans and their reverse reflection on social interactions of its inhabitants (Douglas, 1983). The environment is not a unidirectional phenomenon consisting only of natural and man-made features; rather, it additionally entails the counter-response of inhabitants to these features, in terms of psychosocial behaviour, mental maps, associations, memories and attachments to the surroundings. In this context, the term, 'front-stage of the city' refers to the set of all visual elements that city dwellers experience and which may influence them. These include the streets, poles, buildings, scenes, commercial areas, outdoor advertisements and billboards, as well as natural resources like rivers and forests.

In contrast, the term 'back-stage of the city' refers to the dwellers' behaviours, social life, walkability and interactions with the front-stage. Consequently, the term 'urban aesthetics' is not only limited to beautiful, grand or unique buildings (Fig. 1). Rather, it is the collective impression of the degree to which a space brings harmony, blending, culture, identity and association to its population. The term also encompasses the overall liveability of the city spaces.

The remainder of this chapter serves to set the scene for discussing the subject of visual pollution. In particular, it helps to answer a number of basic questions such as: What is visual pollution? How does it affect urban

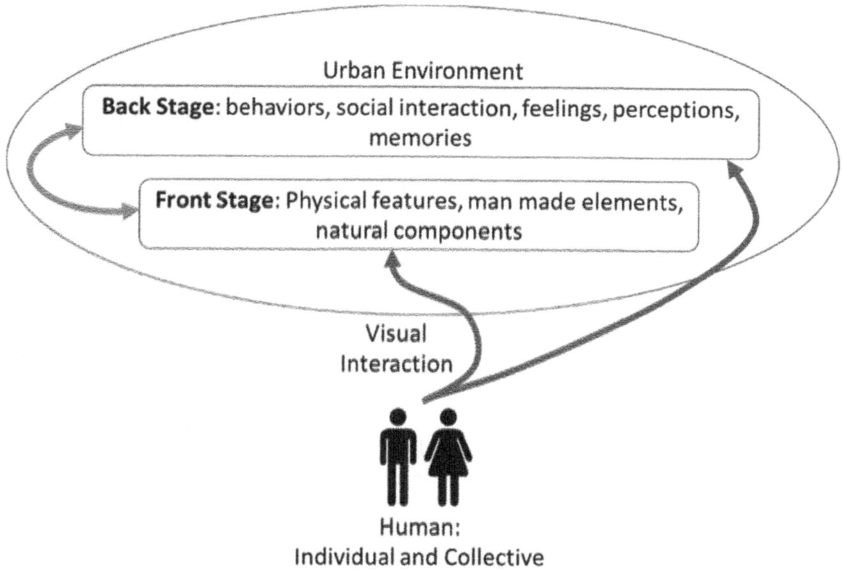

Fig. 1. Human Interaction with Urban Environments.

environments? What are key elements of visual pollution? Why is it important to manage it in systematic ways? How do we address its management? The answers to these questions serve to highlight the importance of the Sustainable Development Goals (SDGs), particularly SDG 11 (Caprotti et al., 2017; UN-Habitat, 2018; Vaidya & Chatterji, 2020), which emphasises the importance of making cities safe, inclusive, resilient and liveable. We discuss the evolution of the concept of visual pollution, its subjectivity, qualitativeness and challenges of defining universal terms. We also discuss the relationship between residents and their surroundings, and how they affect each other. Finally, we explore the elements of visual pollution, including the visual pollutant objects, their characteristics and relationship to different geographic and cultural settings.

1.1 Urban Aesthetics, Harmony and Balance of the Built Environment

The inhabitants of a built environment constitute its central elements; they walk in the city's streets, they observe and experience the surroundings, and they reflect upon these observations and experiences. Their perceptions of the front-stage of their neighbourhood in turn influence the back-stage because social values are affected by the physical and built environment. For instance, if the inhabitants view beauty, they will pass on positive impressions.

Conversely, if they perceive ugliness in their environment, they will pass on negative impressions. The values of coherence, legibility, order and aesthetics of the living environment as perceived by its population have been discussed at length by urban planners, urban designers, psychologists and architects (Nasar, 1988).

Research has shown that all components of the living environment are interlinked with residents as they transmit messages and energy (positive or negative) to their receivers. The urban theories of environment and semiology have quoted such effects repeatedly. Since the mid-twentieth century, the discussion about urban aesthetics has not only been limited to the beauty of a city but has also generated a discussion beyond the city's physical appearance. This paradigm shift has repositioned city entities as messengers of meaning, placing value on their traditions and producing connectivity to their culture. The city has been recognised as a source of identity, with which its residents feel connected. This positioning acknowledges that human beings are 'symbol mongers', and their perceptions are based on the images they observe from their surroundings. Subsequently, these images transform into symbols in their brains and ultimately, a meaning is imparted to them that registers the character and memory about a particular environment (Mennan, 2009; Portella, 2014).

In the early twentieth century, Gestalt's theory of perception and cognition was developed to state that 'the whole is different from the sum of the parts' (Mennan, 2009). According to this, the things that human eyes see in their surroundings make memories. Human eyes 'read' the information around them, and the visual sensors in the human body transmit to the brain, which ultimately builds their memory of a particular public space (Portella, 2007). The principles of perception and cognition explain the relation between the visual character of physical objects in the built environment and human memories of them. These principles have been characterised as *continuation, similarity, closure, proximity* and *figure* (Condly, 2003).

There began to be a realisation that urban planning must move beyond a consideration of the physical structures alone and instead must take into account the diverse elements that affect the residents of an urban environment. Professionals, urban planners and psychologists inferred that all elements of a neighbourhood present images to residents or passers-by. The new body of knowledge became known as symbology, which considers the importance of order and coherence in the urban environment (Langer, 1953).

Regarding the earlier discussion, the move towards aesthetic value, identity and three-dimensional character of a city was initiated during the twentieth century (Xiangzhan, 2008a). For the majority of the twentieth century, Anglo American art philosophy brought out the highest intellectual and creative inputs. However, concerns about environmental aesthetics were acknowledged and raised in the second half of the 1960s. Beardsley introduced the concept of aesthetic justice of the city in 1982 (Hastaoglou-Martinidis, 2011; Mattila, 2016; Mowla, 2011).

As a result, town planning practices were reviewed to incorporate the artistic principles of civic and public art (arte publica) that advocates for 'genuine art of space' all across Europe. Camillo Sitte insisted on transforming the engineering behaviour of town planning towards the aesthetics and beauty of public spaces (Hastaoglou-Martinidis, 2011). Other practitioners of the time were also involved in the debate and recommended similar modifications to the urban environment. These practitioners included Charles Buls in Belgium; Raymond Unwin and Thomas Mawson in England; and Werner Hegemann, Albert Brinckmann and Hermann Jansen in Germany (Hastaoglou-Martinidis, 2011).

Later, in the 1960s, Kevin Lynch introduced the micro and macro perspectives of city aesthetics in his famous book, *The Image of the City* (Lynch, 1960). He characterised the visual qualities of cities in terms of their legibility, structure, identity and imageability. He divided the physical city environment into five categories, i.e. districts, pathways, edges, buildings and nodes. These categories are the anchor points of visual stimuli experienced by the population, allowing them to build memories by reflecting upon the culture, identity and character of a place, rather than seeing them as mere structures of plaster and other materials.

In the last quarter of the twentieth century, many other concepts of urban aesthetics were introduced in the context of the broader function of urban environment, such as aesthetic justice by Beardsley, 1982 (refashioning of the city), human-scale development, walkability and sensory arrangements of the city (Beardsley, 1982; Mattila, 2016; Stamps, 2000; Xiangzhan, 2008b).

The above discussion reveals that urban planning practices and environmental perceptions have continuously evolved. The liveable spaces are now expected to be more than a complex net of lines and multi-options. The quality and life of cities and urban environments are evaluated against beauty, order, visual appeal, cleanliness, aesthetics, sustainability, safety and resilience. Therefore, the design and mutual coordination of physical infrastructures in the built environment are pivot points to engrave hard and soft agendas of sustainable urban places.

1.2 Urban Environments and Sustainable Development Goals

Aesthetics, also known as the philosophical study of beauty and taste, investigates beauty and symmetry as the natural instinct of humans (Kelly, 1998; Kieran, 2005; Townsend, 1997). All living objects, including humans, plants, animals and birds, have an innate preference towards beauty and cleanliness. Evidence of this in humans is observable as far back as the ancient Indus civilisation, which was designed on a grid-iron pattern. This was a huge development, consisting of 1,500 big and small cities, each having a uniform system of urban planning in which even the house and brick weight, size was standardised. Covered drainage and underground storm/sullage water systems were also introduced (Mishra, 2014). It is one of the basic human rights to see the beauty around them, just like the rights of good food, clean air, education (Johnson, Symonides, & Unesco, 1998; Morsink, 1999). Accordingly, it can be safely concluded that visual appeal, discipline and beauty are biological impulses of humans, whose imprints can be charted across human history.

Referring to the definitions introduced earlier, living places (either urban or rural) can be viewed as environments in which the inhabitants, physical objects/structures, natural features and visual elements are the 'ingredients that interact with one another'. Among these elements, cities cannot be designated as static objects, since their character is constantly changing; this is due to a plethora of push-and-pull factors such as migration, degradation, change of economy, infrastructure demands, technology inclusion, disasters and pandemics. The latter half of the twentieth century has witnessed a vast escalation in urbanisation trends. In 1900, only 10% of the world's population was urban. However, within a period of 40 years during the latter half of the twentieth century (1950–1990), the urban population increased from 200 million to over 2 billion. i.e. a ten-fold increase. Furthermore, it is estimated that two-thirds (68%) of the world population will be urbanised by 2050. Therefore, the current era is known as the Anthropocene and urban age (Anderson, Okereke, Rudd, & Parnell, 2013). Although urban population growth has directly contributed towards to the growth in the global gross domestic product (estimated 85% GDP), this growth also poses threats and challenges to healthy urban environments. The rapid population shift from rural to urban, alongside rapid urbanisation, has transformed urban areas into consumerist giants. The rapid increase in the number of urban inhabitants is leading to greater use of resources, greater generation of waste and, as a result, larger amounts of pollution (Hakiminejad, 2018).

Over time, concerns about the quality of life in cities have been increasing. In response to this, the United Nations has established a dedicated goal concerning the quality of settlements. The SDG 11 aims to make cities and human settlements inclusive, safe, resilient and sustainable by 2030. This agenda has been endorsed by a total of 193 countries around the globe (Devisscher, Konijnendijk, Nesbitt, Lenhart, & Salbitano, 2020; United Nations, 2015; Sachs, Kroll, Lafortune, Fuller, & Woelm, 2022). The ever-increasing rural–urban migration has shifted the frame of urban governance and introduced the SDGs. Visual aesthetics, as a whole, is a rarely considered aspect of urban planning, even though it constitutes an important means of achieving safe and liveable cities (Russell, 2018).

Similarly, while cognition theory advocates for simple and clean surroundings, the overcrowded visual display of information creates disorientation, visual chaos and is difficult to digest. It has been argued that enhanced visual quality of space is related to increases in the safety and behaviour of residents, and to better communities at a larger scale (Amin, 2002; Portella, 2014). In short, a visually clutter-free, harmonious and uniform environment is a key factor in ensuring sustainable communities.

1.3 The Emergence of Visual Pollution: A Historical Perspective

The word 'pollution' emerged in old French and originates from the Latin term 'polluere' that means 'to soil or defile'. Noah Webster's first dictionary defined pollution in terms of five different categories: general, religious, Jewish economy and medicine. All of these definitions collectively viewed pollution as different ways in which human activities can affect their lives and/or their environment. Air and water pollution were first acknowledged after the Industrial Revolution, which was also the initial trigger for urbanisation. In response to the challenges of rapid rural–urban migration during the nineteenth century, US laws discussed both types of pollution, due to their harmful influences on the human environment. Visual and light pollution were generally acknowledged within the same period (Nagle et al., 2009). However, the larger number of references to air and water pollution in the literature before the twentieth century (Nagle et al., 2009) indicate that visual pollution was explored later. It primarily emerged and began to be researched in the mid-twentieth century. In the 1970s, a number of new pollution types were recognised and accepted. For example, several laws and treaties on noise pollution were introduced in this decade.

Initially, the treatment of visual pollution was aimed at combatting the nuisance and proliferation of outdoor advertisements (e.g., billboards, hoardings and commercial signage), which began to emerge in the developed world in the 1960s. Rapid globalisation quickly accelerated the spread of ideas and consequently, there was an increase in the volume of these advertisements. In response to this, many rules, acts and policies were introduced to protect, preserve and enhance the urban visual environment (Wakil, Hussnain, Tahir, & Naeem, 2016). An example was the Highway Beautification Act of 1965, which resulted in four states in the United States (Vermont, Alaska, Hawaii and Maine) banning billboards in the second half of the twentieth century. This subsequently led to a rise in levels of awareness about visual pollution across different countries. Various communities began to raising concerns by demanding visual improvements in their environments, while governments responded by showcasing their mitigative actions. In parallel to this, researchers and professionals determined that an urban environment cannot be deemed acceptable merely because it contains several grand and colossal structures; aesthetics should not be limited to the presence of certain types of buildings, but instead should be focussed on ensuring that the environment as a whole is visually appealing to inhabitants, since this acts as a positive source of energy. This point of view is clearly explained in *The Image of the City* (Lynch, 1960).

Furthermore, William Hudson established an Advisory Committee on the Aesthetics of Major Engineering Structures, which was the first initiative of its kind (Maclurcan, 1973). It led to the visual pollution of outdoor advertisements being recognised as a specific issue and consequently, local authorities, urban and town municipal committees had the mandate for its control and management under amenity and public safety principles (Wakil, Hussnain, Waheed, & Naeem, 2016). Later, a paradigm shift was observed in terms of the regularisation of outdoor advertisements, which introduced public–private partnerships and consciousness about the content of these advertisements. Approaches to controlling visual pollution therefore began by attempting to improve the collective visual aesthetics of the living environment.

Despite the introduction of these measures, the proliferation of outdoor advertisements continued to exert a negative impact on visual aesthetics in different parts of the world until the end of the twentieth century and thus became a crucial issue to address. The debate began in the print media, in which McMahon constituted a prominent voice. In particular, debates surrounding billboards and hoardings extended beyond the narrow focus on their marketing use, and also considered their impact in terms of the principles of visual perception of space and urban aesthetics. This debate was taken seriously by many countries and resulted in the introduction of a range of legislative solutions, including the South African manual for outdoor advertising control 1998, the UK clutter code, the transport corridor outdoor advertising and signage guidelines, New South Wales, Australia etc. Likewise, Finland,

Netherlands, Sweden, Denmark, Scotland, Japan and China have all introduced restrictive control policies for outdoor advertisements. Despite the introduction of these measures, visual pollution continued to increase. However, additional measures continued to be introduced, including the Clean City Law in Sao Paulo, which was passed in 2007 to prohibit every form of outdoor advertising in the city. While, on the one hand, this large-scale operation against the visual blight of billboards and hoarding was criticised by marketing agencies, it was also praised by residents and urban planning professionals (Section 3.1.3).

Although there has been a dramatic increase in the awareness of visual pollution in developed countries since the beginning of the twenty-first century, and various measures have been developed to combat it, concerns still remained about developing countries, where visual pollution was still increasing due to weak urban governance. The observations and challenges of visual pollution faced by the developed countries have been charted across a period of more than 80 years. Meanwhile, definitions of visual pollution have broadened to encompass a range of pollutants in addition to outdoor advertisements. As such, there is a strong need to better quantify and assess the issues, and to develop an improved policy and management framework (Fig. 2).

1.4 Definitional Diversity

Owing to its subjective nature (see section 1.5), we can find many definitions of visual pollution in the literature. It has been declared as a separate, distinctive pollution category like air and noise, which is as noxious and harmful as others, and it is simply 'litter on a stick' (McMahon, 1998). Additionally, the US Supreme Court declares that visual pollution: '. . . is not limited to the air we breathe, and the water we drink. It can equally offend the eye and ear' and harms the visual quality of the living surroundings (Nagle et al., 2009). Numerous authors have defined visual pollution as a set of elements that can offend human vision, spatial orientation, physical, mental health and have psychological and economic effects on a community (Choudhary, 2016; Nami, Jahanbakhsh, & Fathalipour, 2016). In addition, all elements which a community finds unattractive, ugly, intrusive or disturbing fall under the category of visual pollution (Muthukrishnan, 2015). Meanwhile, Nami et al. (2016) and Jana & De (2015b) describe visual pollution as an 'unbridled and uncoordinated diversity' of form, colour, light, and materials, and an accumulation of heterogeneous visual elements that make the man-made environment and urban landscape ugly and unattractive. The work of (Chmielewski, Lee,

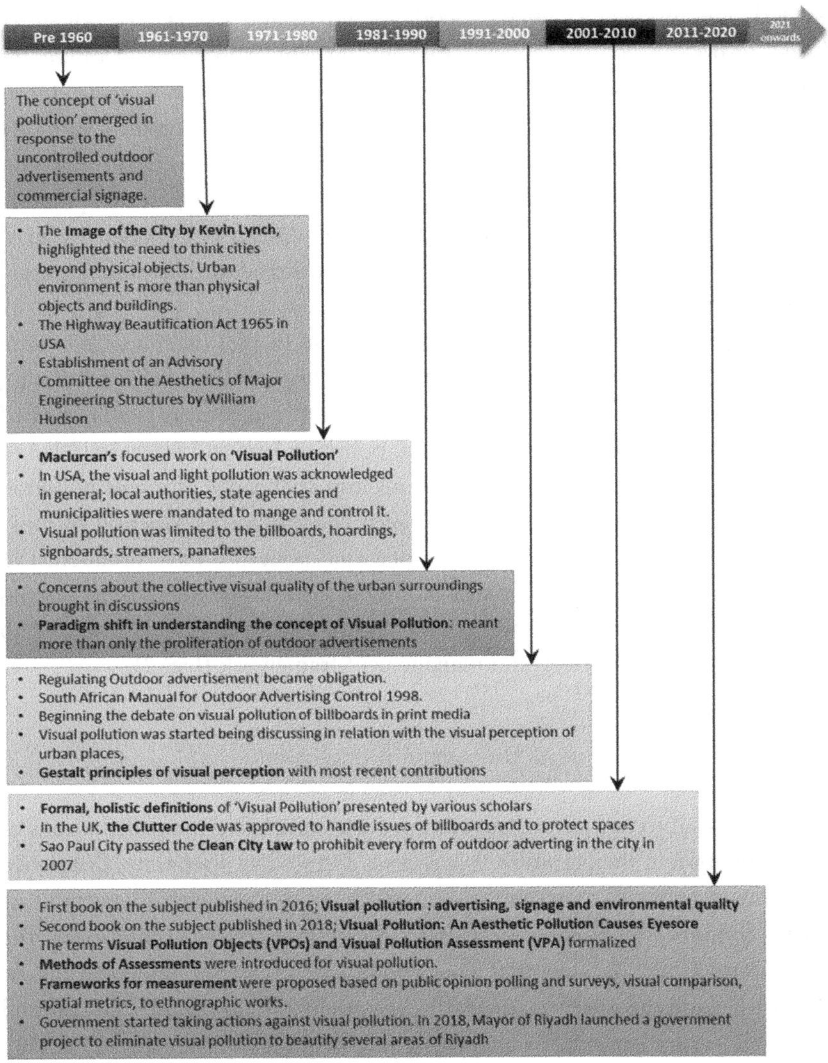

Fig. 2. Historical Emergence of Visual Pollution as a Subject.

Tompalski, Chmielewski, & Wężyk, 2016) defines the term as compound and the resultant effect of 'clutter, disorder, and excess of various objects and graphics in the urban landscape'. Urban visual pollution is the negative physical impact of several objects, which have a direct as well as an indirect relation with the quality of the built environment and which ultimately have implications for humans living in that place (Cercleux, Merciu, & Merciu, 2016; Elena, Cristian, & Suzana, 2012; Galea, Ahern, Rudenstin, Wallace, & Vlahov, 2005; Kamali & Tahmouri, 2013; Parsons et al., 2010). Visual pollution is linked with the visual quality of public and community spaces (Portella, 2014), including regularity, order, beauty, symmetry and simplicity. Furthermore, visual pollution is an accepted phrase commonly used in North, South and Central American countries. All unattractive visual elements of a streetscape, such as billboards, commercial signs, litter, graffiti, telephone lines and poles, generally come under the heading of visual pollution (Cauduro, 1981; Cullen, 1961; Dunn, 2006; Klein, 2000; Minami, 2001; Moles, 1987; Plummer, 2006; Portella & Reeve, 2006; Scenic America, 2000).

The variations in the definitions of the 'visual pollution' reveal that the scope of concept continues to evolve with each new associated research contribution. Nevertheless, there are shared aspects in the viewpoints of most researchers. It is generally agreed, for example, that visual pollution is related to the degree of order, discipline, uniformity, blending, coherence, pleasant appeal and is proportionate to the visual quality of the built environment.

1.5 Characteristics of Visual Pollution

Context: The sensitivity to visual pollution is proportionate to the nature of a person's surroundings and their physical quality of life; the benchmark for the definition of beauty will be different for a person already living in a planned, organised and appealing place, compared to a person who is used to living in an unplanned and cluttered environment. The spread of visual pollution is associated with multi-social indicators (tested in some parts of the developing world). It may be that not all inhabitants of the same urban environment have the same attitude towards visual pollution. In considering the subject of visual pollution, it is important to take into account the character of a place and its function. In particular, in order to explore the extent of visual pollution in an area (Fig. 3), it is vital to consider the level of urbanity of the environment (urban, peri-urban to rural), the dominant type of land usage, the socioeconomic character and whether the place is in a developed or developing country.

Fig. 3. Visual Pollution in Different Geographical Contexts; the Clutteredness of Hanging Wires Increases from Developed to Developing Countries. The Sensitivity of the Residents to Clutter is Also Relational. Braidwood Australia; Lahore, Pakistan (Left to Right).

Subjectivity: Visual pollution involves humans, and every researcher and person have their own social, cultural and geographical backgrounds. Therefore, perceptions of the front-stage of the city vary from person to person (Davies & Mah, 2020). Moreover, unlike air, water, noise and land pollution, it is only recently that efforts have begun to develop standardised, quantifiable tools to assess the levels of visual pollution, which will to quantify the results of visual pollution.

Complexity: The variety of pollutants (visual pollution objects [VPOs]) and their intensity vary in different contexts. There are many variations in the expected individual and community perceptions of the urban space and its relationship with diverse VPOs; since these objects are managed by diverse stakeholders, visual pollution is a complex subject to engage with. This complexity can discourage policymakers from instigating mitigatory actions (Fig. 4).

The emerging area of research: Up until 2020, fewer than 300 research publications containing the keyword 'visual pollution' had been published in the English language; the majority of these works were published in the last decade. However, research into the subject is rapidly expanding, and researchers are increasingly investigating its less explored dimensions. Over time, the focus has shifted from conceptualising and describing visual pollution to identifying VPOs by employing artificial intelligence and machine learning. More recent works have focused on the development of assessment frameworks and the introduction of control measures.

1.6 Visual Pollution Objects

Physical features are one of the key components of the built environment that are also characterised as the front-stage of the city (see Section 1). Some of

Fig. 4. A Contrasting View on the Presence of Complex and Different Character of Visual Pollution from Developed and Developing Countries. Braidwood, Australia; Rawalpindi, Pakistan (Left to Right).

these features are considered to be inappropriate in their urban surroundings; the imbalanced treatment of human interaction with these elements and their respective visual output create visual pollution as perceived by the human eye. VPOs correspond to all physical objects that are considered to be 'pollutants', i.e. objects that create visual disturbance and contribute to visual pollution. In an urban context, we specifically define the term 'VPOs' to refer to all man-made features and their physical characteristics (placement, appearance, size, colour, view and functional hindrance etc.) that affect the visual quality of the urban surroundings in terms of how they are perceived by the human eye.

The term VPO is derived from visual impact assessment studies of natural landscapes. In these studies, scholars use it to refer to physical components that can decrease the landscape's visual quality, contribute towards diminishing visual significance or obstruct the view of valuable natural complexes (Kamičaitytė-Virbašienė, Godienė, & Kavoliūnas, 2016). A similar analogy is valid for urban visual pollution. It has been reported that VPOs, either individually or in combination, can impact upon human health (Jackson, 2003; Tobergte & Curtis, 2012), distract drivers, particularly along main highways (Dukic, Ahlstrom, Patten, Kettwich, & Kircher, 2013; Edquist, 2009; Elena et al., 2012; Jana & De, 2015b), reduce property values, deface public places, spread annoyance, encourage needless consumption or affect the identity of places (Nagle, 2009; Wakil, 2012; Wakil et al., 2016).

An important question is how to determine whether an object should be considered as VPO. This can be established by assessing the extent to which the object complies with the criteria listed below:

- Visually disturbing/annoying for passers-by
- Non-uniform within any existing urban settings
- Out of scale with respect to the surroundings (size, placement)
- Against the principles of beauty, safety and health (physical + mental)
- Hinders the ability to carry out routine tasks of urban life
- Has the potential to challenge public safety
- Indecent content displays

If any physical element within a particular urban environment complies with one of the points aforementioned, it may be classified as a VPO. It should be noted that a specific object that is considered a VPO in one environment may not necessarily be a VPO in all environments; objects may be perceived differently according to the precise characteristics of the urban environment in which they occur.

A list of VPOs has been prepared for this book from various sources, including a literature review, personal observations, a public survey and a university-based photo competition (Fig. 6). A study of the literature reveals that the different types of pollutants which contribute to visual pollution include: dilapidated old buildings, unkempt properties, litter, trash, heaps of garbage, over ground power, transmission and communication lines, graffiti, weeds and intrusive outdoor advertisements (Bankole, 2013; Jana & De, 2015b; Muthukrishnan, 2015; Nami et al., 2016). All of these elements contribute towards the degradation of the built and natural environment and its ambience. It is clear that outdoor advertisements, billboards, hoardings, posters and banners are the most ubiquitous among all identified VPOs (Carrizo, 2016; Jana & De, 2015b). A study shows that among all pollutants of visual pollution, billboards constitute 34% of VPOs (V.A. Gokhale, Mayuri Raichur, n.d.). Outdoor advertisements are of serious concern because of their inappropriate placement, inappropriate height, colours, contents, excessiveness in amount, energy consumption, huge structures etc. (Choudhary, 2016; Jana & De, 2015b). Furthermore, billboards are the biggest threat to urban communities' distinctive and unique character and periphery, due to their high proliferation. Consequently, they have become known as 'visual pollution', 'sky trash' and 'junk mail of American roadway' (OCE, 2014). They have infiltrated every location with their various kinds, including political, property, entertainment, brand boards and have changed the character of public spaces into commercial ones (Molina, 2006; OCE, 2014). 'Simply stated, "blight" is something that impairs or destroys the appearance and results in a deteriorated condition. In recent times, urban blight has come to include a wide range of visual pollutants that degrade the ambience of our communities (Jana & De, 2015b).

Box 1

Visual Pollution Objects Identification Process

Only a few VPOs have been frequently listed in the literature, including outdoor advertisements and billboards (Fig. 5). However, through a public survey, 'Does "Visual Pollution" Exist?' (Available at https://goo.gl/forms/LjKobwAK9m1-wUBZc2), 107 participants were asked to list what they find visually annoying in their neighbourhood or urban fabric around them.

Similarly, a photo competition was arranged among urban planning students to identify and capture VPOs in their surroundings. In this task, an open photo competition was announced at the University of Engineering and Technology, Lahore, where undergraduate students were asked to take pictures of any object/phenomena that they felt to be annoying or visually pollutant at their workplace or in their residential surroundings. The announcement of this competition was made through a printed poster and a university-based social media group. In the poster, the subject's background was explained, and potential pollutants were also listed as examples. Participants were encouraged to capture photos of their surrounding urban fabric, which offend their eyes and create a visual nuisance. Detailed information about the submission specifications was also provided.

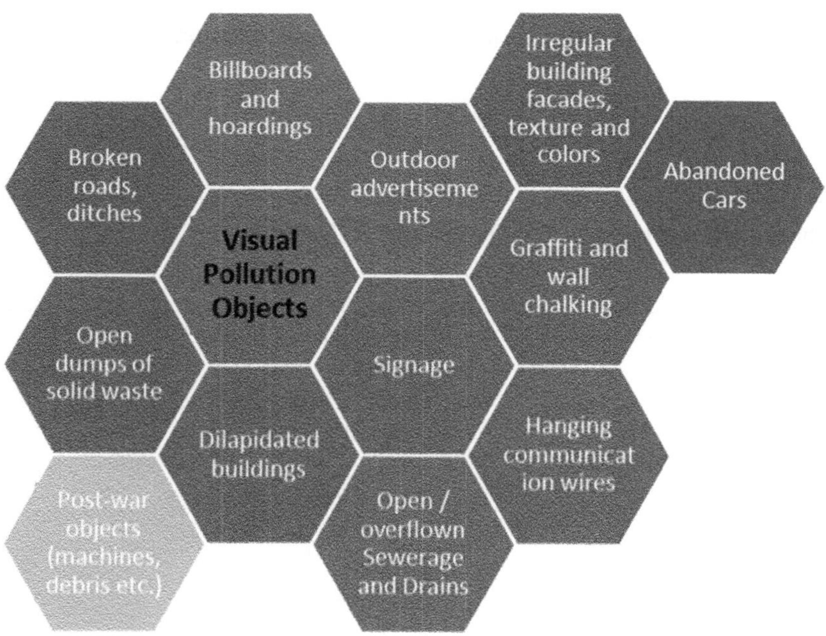

Source: Wakil et al. (2019).

Fig. 5. Common Visual Pollution Objects.

In addition to the literature survey, a detailed study was undertaken to discover additional VPOs. The first stage of this study resulted in a long list of 42 VPOs, including several localised and non-conventional VPOs, e.g. hanging wires, electricity transformers, broken poles, dilapidated buildings etc. Previous studies have not considered such types of VPOs in detail. A panel of experts classified the collected list of VPOs based on similarities between them, which resulted in 10 broad groups, as presented in Table 1. Several images

Fig. 6. Word-Cloud Representation of the Visual Pollution Objects from Literature.

Table 1. Classification of VPOs in Major VPO Groups.

VPO Group	VPOs
1. Outdoor advertisements	• Bright digital signboards
	• Billboards
	• Political advertisement
	• Unethical advertisements
	• Banners
	• Hanging steamers
	• Irregular signboards
	• Posters
2. Graffiti/wall chalking	• Wall chalking
	• Graffiti
3. Open dumps of solid waste	• Solid waste dumps
	• Medical waste
	• Overflowing trash bins
	• Construction material dumps
4. Overflown sewerage/drainage	• Uncovered manholes
	• Blocked manholes
	• Standing water in streets
	• Open sewers
	• Wide-open drains

Table 1. *(Continued)*

VPO Group	VPOs
5. Architecturally poor structures	• Bad building structures, out of proportion • Irregular building faces • Uneven and irregular building skyline • Bluewater tanks on rooftops • Dish receivers and broken antennas
6. Dilapidated buildings	• Dilapidated buildings • Squatter settlements • Poorly maintained structures • Unpainted buildings
7. Encroachments (temporary and permanent)	• Roof projections in front of houses • Uneven ramps • Encroachments (permanent and temporary)
8. Various poles and transformers	• Improper placement of electricity poles • Communication towers • Improper placement and broken/leaning telephone poles • Improper placement and broken TV cable poles • Improper placement and broken streetlights
9. Hanging and cluttered wires	• Cluttered Electric, communication and other wires
10. Broken roads/ditches	• Broken roads • Ditches

illustrating key VPOs found in the streets of Lahore, Pakistan, are presented in Fig. 7. More graphical evidence of these VPOs can be accessed at https://urbanvisualpollution.wordpress.com.

1.6.1. *Outdoor Advertisements*

Billboards/hoardings, streamers, banners, posters and signboards come under this category. These objects are considered as VPOs because of their inappropriate and varying sizes, unstable structures, proximity to land uses (especially educational, religious and residential), visual and functional hindrance, content, colour, worn-out physical appearance and improper placement.

a. Open dumps of solid waste

b. Overflowing waste bins

c. Wall chalking

d. Graffiti

e. Cluttered electricity wires

Fig. 7. Visual Pollution Objects and Their Presence in a Developing Country Context.

1.6.2. Graffiti/Wall Chalking

Graffiti consists of illustrations in paintings, drawings and sprayed presentations in public spaces. In contrast, writings on public walls without permission are called wall chalking. Both graffiti and wall chalking fall under the category of VPOs because of their unwanted presence, along with their controversial and debatable content.

1.6.3. Solid Waste

This visual pollution group includes dumps of solid waste, waste bins that are overflowing, medical waste and construction debris that affect urban aesthetics and functions due to their presence, improper placements, sizes and smells.

1.6.4. Sewer and Drains

Sewerage manholes, stagnant water on the street, open drainage and sewerage drains are all classified as VPOs. Furthermore, open and choked sewerage connections and drains are a source of functional impediment and danger for the public because of their smell and blockage.

1.6.5. Architecturally Poor Buildings

Buildings which are out of scale or not in keeping with their surroundings cause visual displeasure. Other types of visual pollution falling under this category include uneven building skylines, unattractive and heterogenic front elevations, projections that hang out of the property line, extended ramps, irregular and contrasting installation of ventilators and windows in building facades.

1.6.6. Dilapidated Buildings

Derelict and decaying building structures are also a cause of visual pollution because of their shabby appearance and crumbling nature. They affect the visual beauty of the living environment and pose serious threats to adjacent properties and humans.

1.6.7. Encroachments

Encroachments are defined as the illegal occupation of public places in markets and streets. Encroachments hinder daily activities and movements by blocking thoroughfares. They also affect the visual accessibility and clear vision of public areas. This group of VPOs includes extended projections and ramps in front of built areas and in public spaces and rights of way, along with permanent and temporary structures and hawkers in public spaces.

1.6.8. Poles and Transformers

Telephone and mobile masts, electricity poles, transformers and streetlights constitute key VPOs when they are placed in dangerous locations that are close to the mainstream activities. Issues with them include varying heights, cluttered appearance, unlimited counts, dis-segregation of different poles, torn-off condition and structural instability; these can all create visual chaos and functional barriers in urban localities.

1.6.9. Hanging and Cluttered Wires

This includes electricity, communication and cable wires when they are creating a visual and functional nuisance. They may be jumbled, hang low in front of residences, commercial hubs and shops and on streets, posing threats to passers-by and citizens and producing visual blight.

1.6.10. Broken Roads

Uneven road surfaces, ditches and pothole-ridden streets constitute a further dimension of visual pollution that influences the smooth traffic flow on the roads. Such issues can cause road accidents and vehicle damage. Standing rain water in ditches and overflowing manholes in the middle of roads can also generate epidemics.

1.7 A Multi-Dimensional Challenge: Research, Legal, Enforcement

Since the subject of visual pollution is evolving, it presents multi-dimensional challenges to be handled by urban planners, architects, city managers, government agencies and the public. On the one hand, there is a wide research gap in terms of the conceptualisation and description of visual pollution, VPOs, and their assessment, measurement and quantification. On the other hand, there are regulatory constraints to ensure effective mitigation and control mechanisms. Another tier is the enforcement by diverse stakeholders, ranging from individuals to communities and government and private organisations. The increasing societal demands concerning quality of life, coupled with insensitive community behaviours and attitudes towards visual nuisance, coupled with a contrasting competency of urban governance and deficient policy directives makes it challenging to answer the basic question of 'where to start?'. Chapter 2 of this book presents an in-depth discussion on the current state of knowledge. Meanwhile, the challenges of quantification are addressed in Chapter 4, and mitigatory actions are described in Chapter 5.

1.8 Conclusion

The concerns raised by Lynch in *The Image of the City* relating to the form and impression of a city for its inhabitants are one of the key dimensions of urban planning, urban design and urban environmental studies. It is clear that there are two aspects of city life: front-stage and back-stage. The front-stage constitutes what is experienced by passers-by, either forcefully or willingly, as they move through the city. In contrast, the back-stage concerns the behaviours and social interactions of the city's population, which have an impact on what the city showcases in the front-stage of its neighbourhoods. It is proven that what we see is reflected in our memories. The legibility and imageability of living environments are of key importance in this context. Therefore, visual pollution and urban aesthetics are important agendas to be addressed in the living surroundings of the twenty-first century. Moreover, visual pollution is of great concern, particularly for developing countries, where matters related to the visual quality of the built environment have lower priority than in developed countries.

1.9 Summary

(1) Visual pollution has been explored as a new type of pollution in the man-made environment since the second half of the twentieth century. The research article published by Macluran foresaw it as one of the most

important visual concerns of urban environments that needs to be handled effectively.

(2) The definition of the environment is complex. It is the collective interaction of living organisms, physical objects, natural elements and social interactions. Physical objects in any urban setting constitute the front-stage of the city, while the mutual behaviours of the residents are defined as the back-stage of the city. The back-stage has a direct impact on the front-stage.

(3) Aesthetics is a human instinct and is directly or indirectly affected by the visual quality of their living surroundings. Humans are symbol mongers, which means what they see affects how images are constructed in their brains and how memories develop. The relations between the quality of the living environment and human social life and behaviour have deep roots and have been addressed in research for centuries.

(4) Kevin Lynch presented an alternative framework for the components of the urban environments in his book, *The Image of the City*, which introduced the importance of the visual quality by defining five distinct elements, i.e. nodes, districts, pathways, edges and landmarks.

(5) The work of Lynch and the Gestalt principles of semiology inspired a substantial body of researchers to investigate the need for urban aesthetics and its impact on the population of urban neighbourhoods. It has been found that the character of a city is determined not only by its physical buildings but also by how the urban environment communicates messages to the city dwellers. Whatever the city represents is translated into its residents' minds responsible for the lasting images and impressions of a particular place.

(6) In today's urban age, due to aggressive urbanisation, cities face challenges of visual appeal and aesthetics. The SDG 11 focuses on establishing sustainable cities with visually clutter-free societies.

(7) Visual pollution as a term has been debated since the mid nineteenth century. In more recent times, a paradigm shift has been experienced from the conceptualisation of visual pollution to its effective management. Visual pollution is defined as the visual clutter of many objects that hinder the ability of the human eye to perceive harmony and beauty in an environment. It can also be defined as the negative and imbalanced consequences that result from human activities in the natural environment.

(8) Visual pollution is a relatively novel, subjective and intricate branch of knowledge in which context is of prime importance. For example, the level and type of visual pollution, along with the objects that are responsible for it, will vary according to the precise characteristics of the environment, e.g. whether it is rural or urban and whether it is in a developed or a developing country etc.

(9) The pollutants (i.e. elements of visual pollution) have been termed as VPOs in this book. A comprehensive list of over 40 VPOs has been documented in the chapter.

(10) There are seven indicative characteristics of objects that can be used to determine whether they constitute VPOs. These include being non-uniform or out of scale with their surroundings, having indecent content and being visually disturbing.

(11) Certain legal, administration, research and development areas are required to deal with visual pollution effectively. A large amount of further work is needed to explore the concept visual pollution comprehensively as a new type of pollution.

2

STATE OF THE KNOWLEDGE ON VISUAL POLLUTION

1. INTRODUCTION

Visual pollution is a new type of pollution that has only been recognised in the recent past, as mentioned in chapter one. All around the world, researchers, planning councils and academia are investigating its many aspects, experimenting with and revising techniques to handle it effectively, and formulating strategies to ensure its better management and control. This chapter follows on from the first chapter by presenting a detailed discussion about the current state of the published literature on the subject (books, journals and conference papers) by employing a systematic review. The scoping review on the body of literature about visual pollution has been undertaken using keywords and semantic searches on the major global digital databases. This section identifies the contributions of many researchers. It highlights the knowledge gaps, including definitions and explanations of basic terms, assessment of levels of visual pollution, content and execution of regulatory frameworks from each dimension for further interventions. The literature review reveals an increase in publications on the subject in the last ten years, which constitute an important basis on which to build more advanced knowledge about visual pollution in the future.

1.1 Systematic Literature Review

We adopt a systematic scoping review approach to gather and review previously published literature. Our approach consists of four phases: identifying important research databases, repositories and digital libraries; determining and finalising the research keywords and strings; designing the inclusion and exclusion criteria to identify relevant literature and analysing of the searchable published material. The first phase involved the shortlisting of international digital databases including 'Google Scholar', 'Microsoft Academic', 'Web of Science', 'Art and Architecture Complete', 'Urban studies abstracts via EBSCO', 'APAFT: Australian public affairs (full text) via Informit', 'ScienceDirect', 'Scopus' and 'SAGE Research Methods Online'. Further to the listed databases, individual sites of highly relevant journals have also been searched to ensure that all relevant literature has been considered. Key relevant journals, publishers and generic websites included Taylor & Francis Online, Elsevier, Emerald Insight, Sustainability, MDPI, Google Books, IEEE, IOP Science, SAGE Publishers, Routledge and Wiley. Furthermore, the archives of various urban planning and architecture-related conference events have also been explored.

The second phase consisted of determining different possible keywords, combinations of keywords, semantic and string searches to explore the subject. In this chapter we wanted to capture the full spectrum of the available published material. Hence, semantic search techniques have been used to fetch results beyond the traditional string matching (Kasenchak, 2019). The identified resources were searched using keywords both individually or in combination using truncation, Boolean or proximity operators. For example, 'visual pollution OR urban visual pollution OR visual pollution assessment OR visual pollution measurement OR visual pollution quantification OR visual pollution mapping' AND 'urban'. Every effort has been made to design search strings such that each database is searched exhaustively for relevant knowledge.

The third phase defines the criteria for inclusion or exclusion of the marked articles. Since the publications are limited in this field, they have been searched since the 1950s, rather than focusing only on the last 5–10 years. The other rules of article search were:

- It must have a minimum of one visual pollution relevant keyword in the abstract or title
- It must be published in the English language
- It must be related to the urban environment. All non-urban fields (arts, medical, image analysis, etc.) are excluded from the study.

The fourth phase is the analysis of the searchable published material that has been made using the software tool named 'Publish or Perish' (Harzing, 2007, 2010) that allowed the tabulation of search results from various sources (mainly Google scholar, Microsoft academic, Scopus and Web of Science) against indicators such as the number of:

- Papers
- Citations
- Cites/Year
- Cites/Paper
- Cites/Author
- Papers/Author
- Authors/Paper
- Cites/Author/Year

From the initial analysis, it was found that Google Scholar offered the most exhaustive list of resources. Hence, the more complex searches on combinations of strings were performed on this resource only. It has been ensured that the literature review covers the following areas:

- Definition urban visual pollution and the key contributors to visual pollution in urban settings
- Methods and tools for measuring urban visual pollution
- Handling subjectivity in stakeholders' perception about causes of visual pollution and their relative importance
- Quantification and spatial mapping of visual pollution
- Impacts of visual pollution on the respective urban fabric
- Best practices on the regulatory and institutional setup for managing visual pollution (Chapter 3)
- Stakeholders involved in the management and control of VPOs

1.2 Journal Literature About Different Dimensions of the Visual Pollution

A search for the term 'visual pollution' in Google Scholar resulted in 308 hits, which were manually reviewed to retain only relevant published material, as

Table 2. Frequency Analysis of the Published Material on 'Visual Pollution' (Nov 2021).

Query Term in the Title	Visual and Pollution			'Visual Pollution'	'Urban Visual Pollution'
Source	Google Scholar	Microsoft Academic	Scopus	Google Scholar	Google Scholar
Papers	248	161	68	308	40
Citations	974	463	329	1437	200
Cites/Year	18.38	8.74	6.85	8.76	1.22
Cites/Paper	3.93	2.88	4.84	4.67	5
Cites/Author	521.3	163.43	329	764.05	92.03
Papers/Author	167.99	101.74	68	204.98	22.45
Authors/Paper	2.09	2.57	1	2.13	2.43
h_index	17	10	10	19	8
g_index	27	20	17	32	13
hc_index	12	8	8	15	7
hl_index	6.57	2.63	10	6.81	3.05
hl_norm	11	7	10	14	5
Cites/Author/Year	9.83	3.08	6.85	4.65	0.56

shown in Table 2. After removing the publications related to non-urban fields (arts, medical, image analysis, etc.), a final list of 173 publications was created for the literature analysis (Fig. 8).

Fig. 9 presents the country-wise frequency of publications about visual pollution, which reflects awareness about it in different parts of the world. The illustrated imbalance clearly demonstrates the need to carry out further research and to raise awareness about visual pollution in different areas of the world.

All of the selected published materials were classified into one of six different categories to facilitate a better understanding about the maturity of research in key areas relating to visual pollution, and to help to identify where research gaps remain. The six categories are as follows (Fig. 10):

(1) Conceptualisation: literature presenting the definition and other conceptual dimensions of visual pollution

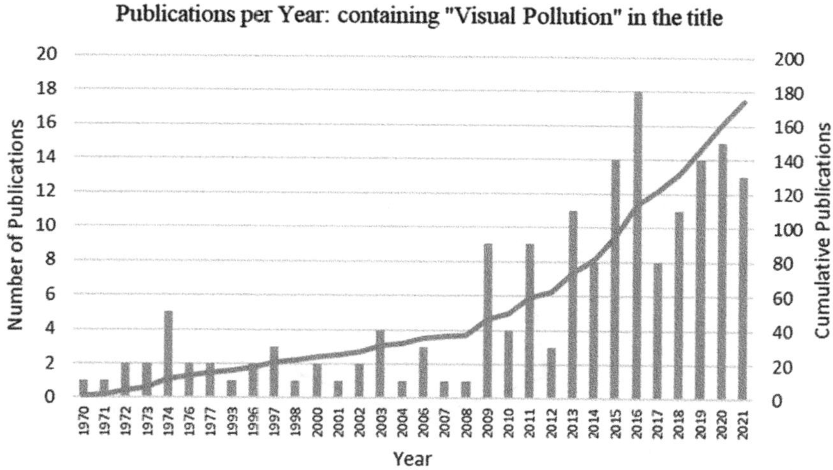

Fig. 8. Publications per Year; Containing 'Visual Pollution' in the Title.

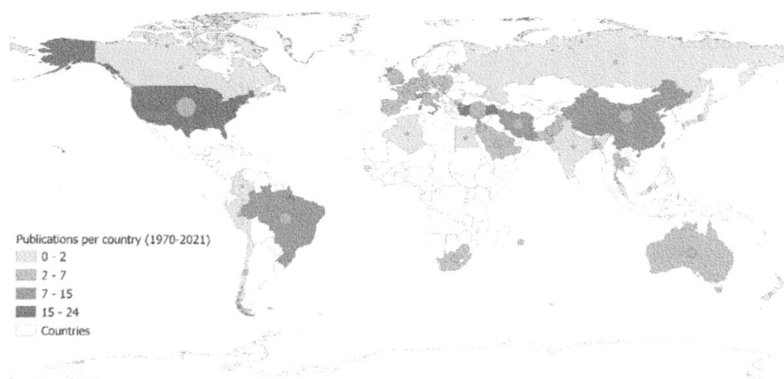

USA	24	France	5	Slovakia	3	Portugal	2	Hungary	1
Turkey	19	Pakistan	5	Spain	3	Qatar	2	India	1
U.K.	15	Poland	5	Thailand	3	Algeria	1	Latvia	1
China	11	Saudi Arabia	5	Bangladesh	2	Belgium	1	Lebanin	1
Brazil	9	Germany	4	Chille	2	Bulgaria	1	Malaysia	1
Italy	9	South Africa	4	Croatia	2	Canada	1	New Zealand	1
Iran	8	Denmark	3	Estonia	2	Colombia	1	Palestine	1
Australia	6	Netherlands	3	Hong Kong	2	Czech Republic	1	Peru	1
Jordon	6	Romania	3	Japan	2	Egypt	1	Russia	1

Fig. 9. Visual Pollution-Related Publications per Country (1970–2021).

(2) Description of visual pollution: literature explaining the phenomenon of visual pollution phenomena and how it manifests itself

(3) Pollution ingredients: literature discussing the various VPOs which contribute towards visual pollution in different places

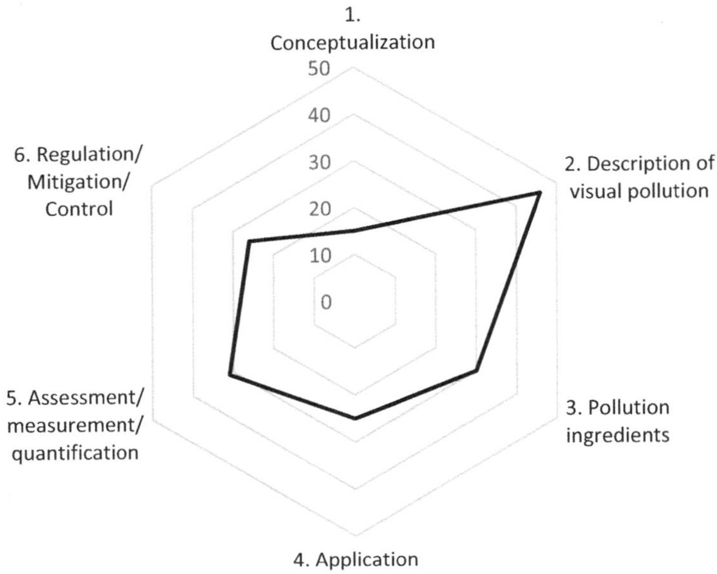

Fig. 10. Content-Based Classification of Visual Pollution-Related Publications.

(4) Assessment/measurement and quantification: literature documenting the ideas and methods of assessment, measurement or quantification of visual pollution

(5) Regulation/mitigation and control: literature showcasing the management, regulation, mitigation and control of visual pollution in urban settings

(6) Application: literature describing the use of different tools and methods to assess one or many visual pollutants with real-world examples.

The results of classifying the published material are presented in a radar chart (Fig. 10). Dimensions with fewer numbers indicate tentative research gaps and highlight the need for potential future work. As shown in Fig. 10, the category with the largest number of publications ($n = 46$) is the *description of Visual Pollution*, followed by *assessment, measurement or quantification* ($n = 31$). Thirty publications are related to *pollution ingredients*, while 26 concern *regulation/mitigation and control*. There are 25 publications related to the *application*, while only 15 are related to the *conceptualisation*.

1.2.1. Conceptualisation

As highlighted in the previous section, the *conceptualisation of visual pollution* is the group with the lowest number of publications. Back in the 1980s, researchers started exploring visual pollution as one of the issues of the urban built environment using various lenses, including art educators (Holdstein, 1974). Similarly, Maclurcan talked about the psychological impacts and changing character of human living places (Maclurcan, 1973). Another group of scholars discussed the establishment of sustainable communities through education and awareness-raising about visual pollution (Dunn, 2006; Kratz & Littlejohn, 2007; Marinescu & Truşcā, 2010). In addition to this, Kumbur and Portella refined the work on the effects of visual pollution on the historic streetscape and urban landscape (Özyurt & Kumbur, 2003; Portella, 2007). More recently, Beer, Sivaramanan, and Wakil have contributed to defining the boundaries of the visual pollution field of study by introducing definitions of terms and highlighting why the human ecology requires the attention of urban policymakers to sustain the visual quality of the built environment (Beer, 2016; Sivaramanan, 2016; Wakil, 2022).

1.2.2. Description of Visual Pollution

This group of publications is focused more on the description of visual pollution. Barrett documented how the national parks in Britain were decaying due to the emerging visual pollution resulting from insufficient moral and financial support from local authorities (Barrett, 1972). Banerjee and Kharate have explored the concept of visual pollution to document its effects on mental health (Banerjee, 2017; Kharate & Banerjee, 2016). Portella has focused on describing commercial signs and outdoor advertisements on the historic streetscape (Portella, 2014), while others, including Anciaes and Blewitt, have reviewed the book of Portella and given their feedback (Anciaes, 2014; Blewitt, 2015). Similarly, El-Ghonaimy has described the visual pollution phenomena and explained the residents' sensitivity to visual pollution in a heritage city in Bahrain (EL-Ghonaimy, 2019). Campos explains visual pollution phenomena from the perspective of social asymmetry caused by the cultural interference by outdoor advertisements (Campos & Cirafici, 2020). Cheng-lin documented the problem of visual pollution in the mountainous and lake areas, destroying touring ecology (Cheng-lin, 2003). Codato has conceptualized the two forms of pollution, audible and visual, occurring in the relationship between society and the environment (Codato, 2014). Correa et al. have explained the visual pollution indicators and their effects on the population in the Metropolitan District of Quito, Ecuador (Correa & Mejía, 2015). Cvetković et al. also see visual pollution as one of the main issues of the twenty-first century (Cvetković & Momčilović-Petronijević, 2018). Some scholars have described visual pollution in different geographical contexts. For example, Szczepańska et al. presented case studies on visual pollution from Poland and Slovakia (Szczepańska, Wilkaniec, & Škamlová, 2019), while Jana documented the presence of visual pollution in Bengal, India (Jana & De, 2015a). Similarly, Xin has discussed the visual pollution in cities of China (Xin, 2011), while Zaydee has described visual pollution in the city of Mosul (Zaydee, 2013).

1.2.3. Pollution Ingredients

The literature review reveals that a group of researchers has contributed to explaining the various visual pollution objects that contribute to visual pollution in different places. In the listing of VPOs, Azeema added billboard advertisements (Azeema, 2015; Azeema & Nazuk, 2016), Bankole talked about Signs and urban graphics (Bankole, 2013), while Dumitrescu and Jensen et al. considered dams, wind turbines and solar panels as a source of visual pollution (Dumitrescu, 2000a; Jensen et al., 2014). Furthermore, Ogunbodede highlighted posters, banners and Billboards as VPOs (Ogunbodede & Sunmola, 2014). Similarly, Fischer declared campus communication as VPO (Fischer & Abllez, 1996); Marks, Nagle, and Roca considered projector phones, cell phone towers and television antennas as visual pollution (Marks, 2009; Nagle, 2009; Roca, 2011). Naguib believed that banners and signage installed on building facades are visual pollution (Naguib, 2016), while Madleňák et al. documented the visual pollution of road infrastructure. Furthermore, Schodorf has listed overhead wires as VPO (Schodorf, 1973).

1.2.4. Application

This group of publications refers to the literature where researchers have applied different tools and methods to assess visual pollutants in real-world cases. Anwer and Aowais have documented the impact of visual pollution on the aesthetics of Eizariya town (Anwer & Aowais, 2019). Emeji has captured the visual pollution in the Nigerian city of Port Harcourt (Emeji, 2011), while McEvoy has presented the case of visual pollution in the Lake Tahoe Basin (McEvoy & Williams, 1976). Similarly, Kadir has documented the visual pollution in the commercial streets of the old city of Mosul (Kadir, 2017), while Kan has written about the presence of visual pollution in Hong Kong (Kan et al., 2014). In addition to that Mendez et al. have talked about visual pollution in public spaces in Venezuela (Mendez & Carmen, 2013). Mousa has documented the presence of visual pollution in the commercial streets in Duhok city (Mousa, 2011) and Sabah has reported on visual pollution In Georgetown, Penang (Sabah & Samad, 2015). Furthermore, Zaeimdar, Yilmaz, and Cvetković have expressed their view about visual pollution in the city of Tehran (Iran), Trabzon (Turkey) and Niš (Serbia), respectively (Cvetković & Momčilović-Petronijević, 2018; Yilmaz & Sagsöz, 2011; Zaeimdar, Sarab, & Rafati, 2019).

1.2.5. Assessment/Measurement/Quantification

From the literature review it is vivid that a proportionally large number of researchers have focused on documenting the ideas on the methods of assessment, measurement or quantification of visual pollution. Some focus on a single VPO, while others view visual pollution as a composite of multiple VPOs. Ahmed et al. have used deep learning to solve visual pollution (Ahmed, Islam, Tuba, Mahdy, & Sujauddin, 2019), while Al-Sharaa et al. have relied on the public survey and cumulative analysis to determine people's thresholds towards visual pollution (Al-Sharaa et al., 2016). Similarly, Allahyari et al. have used SWOT, AHP and QSPM techniques (Allahyari, Salehi, et al., 2017), Atta has applied statistical determination (Atta, 2013), while Bakar et al. have explained the use of cumulative area analysis (Bakar, al-Sharaa, Maulan, & Munther, 2019). The work of Chmielewski and colleagues is more focused on applying geospatial techniques for measuring visual pollution where they have applied intervisibility analysis, citizen science, WebGIS, 3D isovist, and voxel approach and Tangential View Landscape Metrics (Chmielewski, 2021; Chmielewski et al., 2011, 2016; Chmielewski, Samulowska, Lupa, Lee, & Zagajewski, 2018). Zhou et al. have also worked on the same domain and used GIS-based viewshed analysis of visual pollution assessment for the mine environment (Zhou et al., 2011). A significant contribution to this line of work has been made by Wakil et al. who have documented the use of spatial decision support systems and different open source technologies for the assessment and mapping of urban visual pollution (Wakil, Hussnain, Tahir, & Naeem, 2016; Wakil et al., 2019).

Furthermore, Kamičaitytė-Virbašienė et al. have focused on creating the methodology of visual pollution assessment for natural landscapes (Kamičaitytė-Virbašienė & Godienė, 2015). Kucharikova has presented a method for visual pollution localisation through crowdsourcing and visual similarity clustering (Kucharikova & Simko, 2017). Finally another group of researchers has evaluated the impact of visual pollution (Gokhale & Mayuri Raichur, n.d.; Shaban, Suleian, Abdel-Aziz, & Isawi, 2018; Souad, 2014; Xiaolan, Yanyan, & Wenyuan, 2018).

1.2.6. Regulation/Mitigation/Control

This aspect is a relatively less explored area in visual pollution research which focuses on the literature documenting the management, regulation, mitigation and control of visual pollution in urban settings. Abdelhamid has presented the case of visual pollution reduction from Egyptian Cities (Abdelhamid, 2018), Bi-xuan has documented management strategies of visual pollution in historic streets (Bi-xuan, 2014), while Choudhary et al. have also presented a model to mitigate visual pollution by ads and signage (Choudhary & Shrivastava, 2016). Similarly, a group of researchers has documented the regulatory aspects of visual pollution mitigation. In this regard, Bedin et al. have presented the control of visual pollution through municipal law (Bedin et al., 2015), Silva has written about the Clean City Law (Silva, 2020) and Yellin has documented the visual pollution and aesthetic regulations from Los Angeles Bar (Yellin, 1971).

In addition to that, some researchers have presented VPO-specific control measures. For instance, Furze has documented the designs and technologies to reduce visual pollution in wind turbines (Furze, 2002). Eyenga et al. have presented a solution for the shopping bag 'visual pollution' (Eyenga et al., 2002), Jian and Wakil have listed the management solution for outdoor advertisement (Jian, 2020; Wakil et al., 2021), while Rescia et al. have talked about the visual pollution mitigation for offshore plants (Rescia, Corbetta, Puletti, & Rizzi, 2004). Furthermore, Loyola has advocated adopting Berleant's non-cognitive engagement approach as a remedy to visual pollution (Loyola & Ronel, 2018). Similarly, Robinette has proposed to use plants to overcome ugliness (Robinette, 1972), while Sahana focused on the design strategies to reduce the impact of visual pollution in urban areas. A more comprehensive list of studies is presented in Table 3.

1.3 Books on Visual Pollution

Since 1950, only five books whose title includes the term 'visual pollution' have been published in English. The content of three of these books is not publicly available, i.e.:

Table 3. Thematic Details of the Research on Visual Pollution.

Subject	Theme	Key Studies
Visual pollution	Conceptualisation	(Bakhotmah, 1998; Beer, 2016; Dunn, 2006; Enache et al., 2012; Holdstein, 1974; Kratz & Littlejohn, 2007; Maclurcan, 1973; Marinescu & Truşcă, 2010; Meunier, 2010; Özyurt & Kumbur, 2003; Portella, 2007, 2013; Sivaramanan, 2016; Stapleton, 2008)
	Description of visual pollution	(Anciaes, 2014; Babcock, 2015; Banerjee, 2017; Barrett, 1972; Blewitt, 2015; Campos & Cirafici, 2020; Cheng-lin, 2003; Codato, 2014; Correa & Mejía, 2015; Cvetković et al., n.d.; EL-Ghonaimy, 2019; Elena et al., 2012; Flad, 1997; Grygny, 2014; Jana & De, 2015b; Jorge & Gentil, n.d.; Kharate & Banerjee, 2016; Kim & KIM, 2019; Muthukrishnan, 2015; PORTELLA, 2016; Somida, 2015; Szczepańska et al., 2019; Tama, 2020; Tota, 2017; Troy, 1974; Voronych, 2013; Xin, 2011; Zaydee, 2013)
	Pollution ingredients	(Allen, 2017; Azeema, 2015; Azeema & Nazuk, 2016; Bankole, 2013; Dumitrescu, 2000a, 2000b; Ogunbodede & Sunmola, 2014; Fischer & Abllez, 1996; Jensen et al., 2014; Khanal, 2018; Madleňák & Hudák, 2016; Marks, 2009; Nagle, 2009; Naguib, 2016; Nami et al., 2016; Probst, 2016; Roca, 2011; Schodorf, 1973; Silverstone, 1974; Sumartono, 2009; Urban et al., 2020; Zbadi, 1997)
	Application	(Anwer & Aowais, 2019; Benavides, 2017; Córdoba, n.d.; Cvetković & Momčilović-Petronijević, 2018; Emeji, 2011; McEvoy & Williams, 1970; Kadir, 2017; Kan et al., 2014; McEvoy & Williams, 1976; Mendez & Carmen, 2013; Mousa, 2011; Sabah & Samad, 2015; XIU et al., 2009; Yilmaz & Sagsöz, 2011; Zaeimdar et al., 2019)
	Assessment/ measurement/ quantification	(Ahmed, Islam, Tuba, Mahdy, & Sujauddin, 2019; Al-Anbari, Abd, & Obaid, 2020; Al-Sharaa et al., 2016; Allahyari, Nasehi, et al., 2017; Atta, 2013; Bakar et al., 2019; Chmielewski et al., 2016; Chmielewski, Samulowska, Lupa, Lee, & Zagajewski, 2018; Erickson, 1977; Jellet, 1993; Kamičaitytė-Virbašienė et al., 2016; Kamičaitytė-Virbašienė & Godienė, 2015; Khaled, 2009; Kucharikova & Simko, 2017; Lascu, 2015; Melhosseini et al., 2018; Radomska et al., n.d.; Ramos & Treviño; n.d.; Roca, 2016; Shaban et al.,

Table 3. *(Continued)*

Subject	Theme	Key Studies
		2018; Souad, 2014; Gokhale & Mayuri Raichur, n.d.; Wakil, Hussnain, Naeem, & Tahir, 2016; Wakil, Naeem, Anjum, Waheed, Thaheem, & Hussnain, 2019; Wakil, Naeem, Anjum, Waheed, Thaheem, & Hussnain, 2019; Xiaolan et al., 2018; Yao-zhi, 2011; Zhou et al., 2011)
	Regulation/Mitigation/ Control	(Abdelhamid, 2018; Bedin et al., 2015; Bi-xuan, 2014; Cercleux et al., 2016; Choudhary & Shrivastava, 2016; Chuan, 2009; Club, 1966; Eyenga et al., 2002; Furze, 2002; Jian, 2020; Loyola & Ronel, 2018; Nessim, 2020; Rescia et al., 2004; Robinette, 1972; Sahana & Karthigayini, n.d.; Silva, 2020; Stephany, 2010; Yellin, 1971)

(1) 'On visual pollution and the urban environment' by Raphael Engle, 1 Jan 1971.

(2) 'Visual pollution: Beautification programs in the Caribbean' by Judith A Towle, 1st January 1972.

(3) 'Visual Pollution' by Dagda Tanner Mattheus, 6 January 2012. This book is primarily a collection of articles from Wikipedia and other free sources.

(4) 'Visual Pollution; Advertising, Signage and Environmental Quality' by Adriana Portella, 2014.

(5) 'Visual Pollution', by Sivakumaran Sivaramanan, 2016.

A summary of the two books whose content is available is provided below.

1.3.1. Visual Pollution: Advertising, Signage and Environmental Quality, Portella, 2014

Adriana Portella presents her PhD research work in this book. It concerns people's visual preferences and satisfaction regarding the appearance of commercial signs in historic streets of different international cities. The book constitutes the first publication of its kind on the subject. It reflects on theoretical and empirical investigations of commercial signage in the historic streetscape.

There are three main aspects of the discussion, i.e. the effects of commercial signage on historic city centres; the control and management of these signs and the effects of their presence on people's visual perceptions and satisfaction. It combines empirical and comparative studies that include two case studies, from the United Kingdom and Brazil. The author presents her viewpoint using theories, concepts and methodologies from multiple disciplines including urban planning, environmental psychology, urban design and architecture. She concludes by emphasising the need for improved and more specific guidelines regarding commercial signage, in order to ensure that better management of signs can be achieved. Moreover, the inclusion of well-thought-out design principles can lead to more positive attitudes towards commercial signage.

1.3.2. Visual Pollution, Sivaramanan, 2016

The most recent book on visual pollution provides a detailed description of visual and light pollution. This short book covers three main aspects of the issue by providing a short definition, listing the main elements of visual pollution and describing measures that have been used to mitigate them. The main elements of visual pollution are identified as: open dumps of garbage, hoardings, advertisements, curbside wastes, beach trash and various litter, badly maintained or abandoned buildings, curb graffiti, spray-painted names and messages, billboards, street banners, electronic signs and public LED screens, telephone and utility poles, telephone, electricity and television cables and wires, damaged vehicles and buildings, curbside weeds, aquatic weeds and eutrophic lakes, oil spills, curbside potholes, beggars, street children, smog/haze and smoke. A brief account of the effects of visual pollution on the urban environment is also provided. Mitigatory measures covered include various initiatives, including the Highway Beautification Act, Clean City and the removal of billboards in many states to curb the clutter of outdoor advertisements. The author additionally introduces a number of recommendations including the introduction of underground wiring, adoption of the 5S strategy (Sort, Set in Order, Shine, Standardize, Sustain), minimisation of air pollution to combat haze, a ban on open dumping and the encouragement of role plays and guitars on the roadside to discourage beggars. He adds that strict environmental and municipal legislation enforcement is the key to controlling visual pollution. In summary, the book attempts to cover the subject from several different aspects, but in a brief manner.

1.4 Conclusion of the Chapter

Visual pollution has been accepted and acknowledged as one of the most important aesthetic issues of the urban world, which has resulted from inefficient urban governance and a lack of awareness. In international literature, researchers, journalists, urban designers, urban planners and sociologists raise their concerns about this 'visual monster' in terms of its impact on urban environments. Although the subject has already been investigated from a number of various different perspectives, there is still a dire need to carry out further research into this branch of knowledge, particularly to help to improve the quality of urban living.

1.5 Summary of the Chapter

(1) After adopting a rigorous search methodology consisting of four phases, an analysis of existing literature on visual pollution was performed; digital databases were identified and searched, and relevant research works were classified according to six different categories.

(2) The six analysis classes used are: conceptualisation; description of visual pollution; its ingredients; application; assessment/quantification/measurement; and mitigation and control.

(3) The radar analysis shows that most existing research falls under the categories of 'description of visual pollution' and 'application' categories. Meanwhile, the category with the fewest number of publications is the conceptualisation of the subject.

(4) Per year publications and frequency analysis highlight research gaps and areas in which additional future research is most urgently needed.

(5) Only five books whose title includes the term 'visual pollution' have been published since 1950. Among those, three are not readily available.

(6) The books by Portella and Sivakumaran are the most recent works on visual pollution. However, the content of these books is limited in terms of the scope and scale of the information covered. The former book concerns the effects of commercial signs in historic streetscapes and attitudes towards them, while the latter provides a concise account of the theoretical aspects of visual pollution.

3

CASES OF VISUAL POLLUTION MANAGEMENT INITIATIVES

1. INTRODUCTION

The value of urban aesthetics and visual appeal has been recognised as an essential part of the built environment. Since the end of the twentieth century, voices have been raised to promote awareness about this issue. Gradually, in different parts of the world, various actions have been taken to combat the visual nuisance of different sources. Although these efforts have focused on addressing either a single component or a limited set of components of visual pollution, they provide a good starting point from which to explore different institutional and community response dynamics, and to document both effective and ineffective solutions. Such an analysis can provide relevant authorities with the means to take well-informed regulatory actions, which are both cost-effective and justified within the target environment; the nature of such actions is likely to vary depending on whether the target environment is in developed or developing country. This chapter presents case studies of visual pollution management and contrasts the different approaches taken, according to varying perspectives in different parts of the world. Regulatory case studies from the developed world include discussions on the Highway Beautification Act in the United States, the Clean City law in Brazil, the Jacksonville community initiative in Florida and the visual pollution mitigation project in the Kingdom of Saudi Arabia. On the other hand, we also examine management practice stories of countries in the developing world, i.e. Pakistan, Sri Lanka and India.

The practices and regulations are discussed firstly in relation to the countries where they were initiated and regulated. Subsequently, their generic implementation is discussed, to give the reader some insight and effectiveness of these laws as general regulations for introduction in other countries.

1.1 Regulatory Cases

The section presents several iconic initiatives, projects, laws and strategies that have constituted milestones in the history of the management of visual pollution in different regions. The extent of the problem and developed countries' attitudes and approaches are significantly different from those of developing countries. Below, we discuss how different countries have responded to the new urban challenge of visual pollution. The map in Fig. 11 presents the spatial spread of the case studies from developed and developing countries based on the indicators mentioned in Box 2.

1.1.1. The Highway Beautification Act, 1965, USA

Inspiration: In the 1960s, the first lady of the United States, Lady Bird Johnson, pioneered actions against visual pollution on roadways. The Highway Beautification Act 1965 was enacted thanks to her efforts. She believed that, in a vast country like the United States, the purpose of roads extends beyond only transportation and economic growth. Rather, they should allow travellers to experience the nature and beauty of their surroundings. Accordingly, she raised concerns about environmental quality and engineering of roads.

Background: The road network in the United States expanded massively in the first half of the twentieth century, resulting in easier movement of public

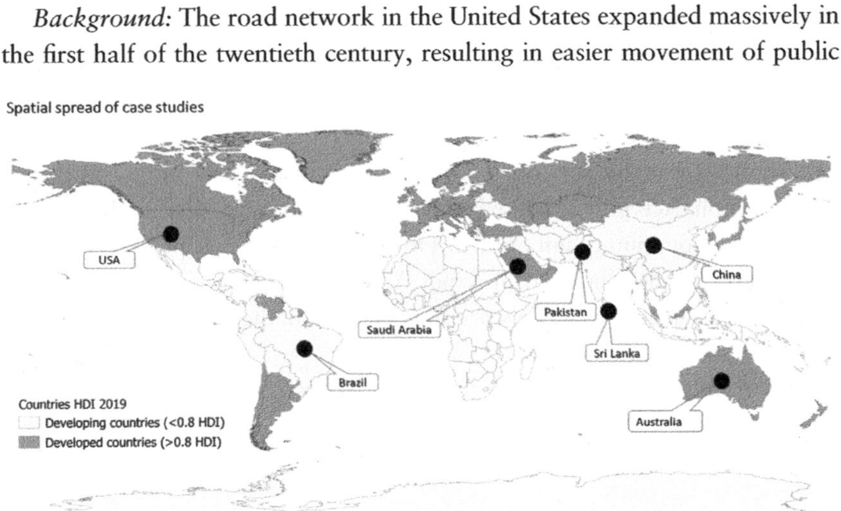

Fig. 11. Spatial Spread of Case Studies Presented in Chapter 2.

Box 2

Gross Domestic Product and Human Development Index

GDP: It is the total monetary value of all products including both goods and services in the market within specific time period of a country. That time period is generally of a year.

The American economist Simon Kuznets developed the current concept in 1934 that was adopted as the key measure of economy of a county in 1944 at Bretton Woods conference (International Monetary Fund World Economic Outlook (October 2021), World Population Review).

HDI: It is the measure of human development in a country on a scale from 0 to 1 by the United Nations. Human development means health, education and life expectancy (World Population Review).

and an explosion of outdoor advertisements, especially billboards and hoardings. The chaos along the country's highways was marked and criticised as 'sky trash', 'the junk mail of the American Highway', 'litter on a stick' and 'visual pollution' by environmentalists. By the end of 1920, the roads were lined with gigantic and monster structures that obstructed views and detracted the attention of drivers. Later, in May 1923, a protest-by-letter against billboards and hoardings was initiated by the Women's Club of the New York States Federation that was supported by other states. They opposed the ever-increasing installation of billboards along highways and demanded that advertisers must confine them to commercial and industrial zones, for the sake of preserving the scenic aesthetics of the United States.

Furthermore, these women sent 4,000 letters per month to the top four advertisers in the country to obtain support for this pledge. Resultantly, the contracts of 14 advertisers were cancelled, and they agreed to remove their billboards from highways. These actions ignited intense interest in the movement. For example, President Johnson became aware of public concern about the marred landscape. In response, he presented a bill in the parliament, approved by the cabinet (Cunningham, 2016), which demanded the removal of outdoor advertisements along the roadside, in order to conserve natural aesthetics and resources (Daluge et al., 2011). As a result, the Highway Beautification Act was enacted in 1965, providing regulations, rules and policies regarding billboards on highways. The president signed the Act in October 1965 and stated that the divine beauty of America would not be disturbed and ruined by reckless human actions during his tenure.

Main characteristics of the Act: The act worked on three areas, i.e. scenic enhancement, junkyard control and regularising of outdoor advertisements. It monitors activities at a distance of 660 ft around interstate and primary highways. However, informative, directional and official signs are exempted from the act. The federal government penalises the non-complying states through a 10% reduction in the states' highway budget.

How it works: The Act does not put a complete ban on billboards; instead, it limits them within commercial and industrial zones. The Act's implementation is associated with city development control, comprehensive plans and zoning strategies to combat the effects of rapid urbanisation in the last nineteenth and first half of the twentieth centuries.

Response to the Act: The Act was high in demand and adopted by the 33 states, followed by zoning regulations that permit dividing a municipality into various sub-districts within districts to formulate different rules for land use, outdoor advertisement and building size. However, the courts did not approve linking land use control with aesthetic purposes earlier. Still, later they vet billboards and outdoor advertisement regulation in the context of public safety, harmony, public health and preservation of property values, which is the only legitimate use of powers (Stensland, 2004). Although, a school of thought also criticises the Act for its expensive administration.

The Act is a great initiative and limits the use of reckless billboards on highways; however, the criticism with regards to its expensive administration and no 'legitimate' linking of aesthetic purpose with land use could cause problem for its implementation in other countries. The Act was administrated in a developed country e.g. America and its use for under-developed countries is still too early to comment, given no proper regulations or regulatory bodies who control the volume and spread of these billboards across highways are present in such countries.

1.1.2. The Jacksonville Community Council Initiatives, 1983, Florida

Inspiration: To make the Jacksonville community clean, visually attractive and beautiful.

Background: The chamber of commerce initiated the programme, further owned by the mayor in 1983, to keep the community liveable. The purpose was to educate Jacksonville citizens and the community to maintain its beauty and handle litter by developing positive behaviours. The program was further sub-divided into projects that worked with the children of the local schools to capacitate and train them about community aesthetics and their responsibilities as a member of society. It raised a sense of ownership for the whole place rather than their homes, the neighbourhood only.

The study was conducted to determine the responses to manage visual pollution of junkyards, signs, billboards, overhead wires and other sources by analysing effects of visual pollution, existing regulatory mechanisms, and improvement of the laws, policies and rules.

Key recommendations: The committee put forward the following recommendations to deal with the serious problem of visual pollution due to the weak commitment of the public official and departments:

- Communities have the legal right and responsibility to maintain the aesthetics of their neighbourhood
- Beautiful, liveable areas bring economic prosperity for it.
- Shaping the ordinances is the first source of refining the spaces
- Improve the visual quality by having more open spaces, underground utilities, tree plantation, monitoring of architectural design, scraping off old signs and likewise.
- Establishment of the visual appearance constituency that can talk about visual issues of the society and can create activities for community visual enhancement
- Restricting grandfathering-in of non-conforming uses
- Different treatment for illegal, on-site, off-site signs and billboards.
- Collaboration with civic organisations, citizens and students (Council, 1985)

The recommendations assume cooperation from the community especially the children within school. This means it will generally work in places where literacy rate is reasonably higher than a certain threshold. The recommendations work only if the cooperation rate is higher than those who neglect or don't accept these recommendations at all. These factors should be considered when implementing it in other sites or countries.

1.1.3. Clean City Law, 2007, Brazil

Inspiration: The clean city law aimed to protect the natural, historical and the cultural environment of the city; enhance, improve and transform the urban landscape and celebrate the beauty of open pathways, unblocked facades and vistas of the buildings.

Background: Sao Paulo is the largest metropolitan of Brazil, South America. With a population of 11 million, it was overcrowded with the flood of 15,000 billboards and other outdoor advertisements, including 16,000 signs, 13,000 metal panels, posters, banners, flyers and streamers that caused visual blight in the city. Mayor Gilberto took the matter seriously and framed 'The Clean City Law' to combat visual pollution. He stated that it is as important to deal with this new type of pollution as water, air, noise and visual pollution are the most conspicuous ones (NYC Global Parnters, 2011). He presented the law bill to the City Council in 2006 that was approved and enforced by the council in January 2007. The business and advertising companies were issued 90 days compliance and removal notice. After passing the prescribed time, all stakeholders, i.e. billboard owners, landowners and advertisers, are charged at a rate of USD 4,500 per day (Silva, 2020).

Main characteristics: The council implements it with zero cost because business owners have to remove billboards on their properties. Otherwise, heavy fines are imposed by the municipality. The mobilisation of the community and awareness meetings with businessmen and store owners helped educate the public for the larger interest of the population and the visual aesthetics of the city.

How it works: It regulates outdoor advertisements on public and private properties. It bars the installation of all forms of outdoor advertisements at towers, lakes, rivers, public places, reservoirs, parks, squares, city properties, trees, motor vehicles, residential zones, light poles, water and gas ducts, watersheds, historical and artworks, traffic signs (Chmielewski, 2020).

Important international recognitions: The law has been acknowledged with the following international awards:

- Expo Shanghai 2010, China
- Emotional Branding Visionary Award 2011, United States
- Werkbund-Label 2012, Germany
- Brit Insurance Design of the Year 2008, the United Kingdom (Silva, 2020)

Although, earlier in history, many states/places have gone through such restrictions, including Hawaii (the 1920s), Vermont (1968), Maine (early 1980s), and Alaska (1998), which have taken similar actions. However, this legislation was the first and unique as it officially declared these visual elements as visual pollution and influenced the world and local authorities like Porto Alegre, Rio de Janeiro and Brasilia to replicate the same measures. Collectively, this law was about the quality of urban spaces and places (Silva, 2020).

Response to the law: The city's inhabitants encouraged and welcomed it, which was confirmed by a public opinion survey held in 2007. Results depicted that 70% of the population was satisfied with the ban on outdoor advertisements. However, legal challenges were faced by the business community, advertisers and landowners, but they participated positively later after repeated meetings of the officials. Another challenge was to shift the public to a new and dramatic city flavour, i.e. a view loaded with outdoor advertisements and then without them (NYC Global Parnters, 2011; Silva, 2020). Additionally, the law is discouraged as it lacks quantifying the visual impact of the visual pollution source. For example, there were no scientific quantification methods to assess the visual pollution of different sources that added biases to the implementation.

1.1.4. Visual Pollution Elimination Programme, Kingdom of Saudi Arabia

Inspiration: The Jeddah municipal authority targeted to make the city an attractive tourist place and sustainable city by reducing the visual pollution spreading in seaport, industrial and commercial areas.

Background: Jeddah is the country's second-largest city, with a population of 4.1 million approximately. It is also the first largest in the province of Makkah which means 50% of the population is concentrated in it as per 2015 estimates. The city's population growth rate is higher than that of the country, i.e., 3.8% per annum. It has been expanded over an urban area of 1,200 sq. km. and is considered a commercial, urban hub that is a gateway to Makkah and Madinah. The city has experienced unprecedented urbanisation primarily from rural–urban migration for better job opportunities and levels of services (Baesse, 2012; Taha et al., 2019). With the changing urban character, the city faced the challenges of the visual quality of the environment. The proliferation of outdoor advertisements, solid waste dumps, damaged and abandoned vehicles, graffiti with time. Therefore, the project for eliminating visual pollution has been initiated under the executive plan for Riyadh environment protection to contribute to the National Transformation Programme by Ministry.

Main characteristics: Compared to the other initiatives, the visual pollution elimination project focuses on different visual pollution objects, including abandoned vehicles. In terms of its execution, the Riyadh municipality was the main responsible agency while the Ministry of Transportation and the high authority of Riyadh development were the participating agencies. The program focused on six key steps to eliminate visual pollution, including the development and placement of standards on the harmonisation of colours in the city environment, placement control on advertisements, removal of elements resulting in visual pollution, setting up a schedule for visual pollution elimination, enhancing the landscapes in the city and providing walking corridors to walkers.

Under this project, the municipality removed the main sources of visual pollution in the first quarter of 2018 and eradicated 12,661 damaged and neglected vehicles from the public streets and city squares. Along with this, the administration cracked down against 15,704 billboards and hoardings within the same time frame. It also includes repainting cluttered public walls (Newsbeezer, 2018).

Response to the project: The project has been scaled up in the capital of the Kingdom of Saudi Arabia. The elimination of visual pollution is the legislative target of the municipality's urban development and environmental plans (Riyadh Environment, 2019). Another debris clearance and solid waste management project was launched in 2014 by the Al-Ahsa Oasis region municipality of Saudia Arabia in partnership with Amarco, Saudi Arabian oil company. The focus was to educate the workers, labour and administration and the citizens to free their area from the visual nuisance of construction debris and piles of solid waste (News, 2014).

An interesting part of this campaign was to first identify the visual pollution objects within given sites and environment. Once the visual pollution objects are identified, it is easier for the government or regulatory bodies to plan and effectively eliminate (or potentially limit) the pollution. Therefore, the countries who wish to pursue this particular campaign should identify the objects which are causing highest pollution and accordingly regulate the elimination of these objects.

1.1.5. Regulating Advertisements: Beijing, China

Visually unpleasant and obtrusive boards of advertisements sabotage the appearance of streets and districts. The city of Beijing is considered among the pioneers within China to regulate outdoor advertising. A commission was formed that reviewed street advertisements' layouts to check visual pollution. The commission was mainly responsible for checking the impact of advertising boards on the streets, both from visual and social aspects, ensuring advertising harmonises with surroundings and evaluating the extent to which advertising signs can distract road users (Center for Local Governance, 2019). The project has been discussed in the news; however, no detailed study is found about the project's outcomes.

1.1.6. Regulating Property Materials: Bundaberg Regional Council, Australia

Bundaberg Regional Council regulates the excess building materials in and around any property under development, and it mandates the owners to maintain the property without impacting the area's visual amenity. The local policy states that it has been made 'to ensure particular building work (involving removal or rebuilding or building work that Council has declared may have an extremely adverse effect on the local amenity, likely amenity, or conflict with the character of the locality) protects and contributes to its surrounding built environment where people live and interact'.

This regulation targets a very specific problem, which may correlate with the previous regulation where identification of the problem precedes the elimination or limiting the visual pollution objects. Since the detailed description of the regulation is not found, it might be hard to implement it in other sites, but the general idea is to involve the community and not completely rely on municipal or government to 'clean' the environment.

1.1.7. Ban on Hanging Laundry in the Balcony, 2008, Kuwait

Inspiration: To reduce visual pollution by hanging clothes on the facades of the balconies of the city's buildings.

Background: On continuous observation of the city front views, the government launched a relentless campaign against hanging clothes for drying purposes on the front elevation of the buildings as they are creating visual pollution and disorder in the regional settlements of Al-Mirqab, Sharq and Bneid Al Qar. Therefore, The Ministry of Municipal Affairs made a legal ban under decision number 190 of 2008.

How it works: The special emergency team executed the decision that conducted regular field visits in collaboration with the Ministry of Municipal Affairs and the Department of Public Cleaning and Road Works. The team sticks a violation sticker on the door if it finds any investigation and imposes fines of 100–300 Kuwaiti dinars. All the citizens and ex-pats must adhere to it (Times, 2020).

This particular issue is apparent in many under-developed countries as well. However, it may not be considered as 'pollution' to small sites e.g. villages or small communities. Enforcing these laws or educating the community will play a crucial role before implementation of such laws in many under-developing countries.

1.2 Cases from the Developing Countries

1.2.1. Pakistan

The subject of visual pollution has been discussed in Pakistan for the last 10 years. A group of urban planners and environmental activists have been raising environmental and social media concerns. Resultantly, a series of temporary actions by the judiciary has appeared in the form of banning billboards in Karachi and removing advertisements in Lahore. Such suo-moto actions were primarily triggered after losing certain lives due to the falling of billboards in rainy or windy seasons. A more structured effort is the Punjab Outdoor Advertisement and Signboards Policy, 2013, which empowers local governments to determine billboards' location, size, type and density. Other similar regulations include Regulations for outdoor Advertisements and Billboards by Parks and Horticulture Authority (PHA), Advertisement, Signage Bye-laws for Karachi 2003, standard operating procedures (SOPs) for Hoarding in Cantonment Areas, July 2011, Military Lands and Cantonment Hoarding Policy 2012 and City District Government Faisalabad Advertisement Bye-laws 2013. Although, the previous review studies critique that Pakistan's regulatory work is more concerned about structural safety and income generation than visual appeal enhancement (Wakil & Hussnain, 2014; Wakil, Hussnain, Waheed, & Naeem, 2016). However, these regulatory steps are still a good example of regulating elements that significantly contribute to visual pollution in an area. Additionally, renovation and rehabilitation projects have occasionally been carried out, especially in major urban areas. A few of those have been quoted as follows:

- Renovation of Gawalmandi food street (Shahzad, 2011)
- An initiative of the urban forest, Liberty Market
- Wall art on the major urban corridors and roads.
- The outcry of environmentalists, journalists and urban planners in print media

1.2.2. Sri Lanka

In Srilanka, Hemantha, an environmental activist and founder of the Centre for Environmental Justice, filed a case against the election posters in 2009 (application no. C.A. 135/2009) and forced the controlling agencies to regulate the display of banners, posters and hoardings. The petitioner highlighted that the education buildings, historical monuments, scenic beauty and visual quality of vistas in the cities had been destroyed due to the proliferation of uncontrolled advertisements. As a result of the case and pressure from communities, Central Environmental Authority drafted the country's Visual Pollution Regulations in 2014, which got approved in 2019 as the Control of BillBoards and Hoardings (Visual Pollution) Regulations No. 01 of 2019 (Hemantha, 2014).

1.3 Lessons Learned; Changed Perspectives in Different Parts of the World

Globally, all countries are categorised into two major groups: developed and developing. This divide is earned based on several factors, including the Human Development Index (HDI), Gross Domestic Product (GDP), political stability, services, infrastructure and industrialisation. Therefore, the countries with a high economy measured by Gross Domestic Product coupled with a good score against the Human Development Index and mature urban and technological development are categorised as developed countries or states.

Since the developed countries are more economically stable and have higher education rates, the level of sensitisation in the general public about visual pollution and communal concern is higher. Interestingly, the review of visual pollution mitigation cases reveals that the efforts in developed countries are more long-lasting and deep-rooted with legislative support than the projects from the developing countries, which are more reactive, temporary and superficial. The former group focuses on eradicating root problems, while the latter group of countries tries to fix the symptoms. Therefore, it is essential to identify the problem first and try to enforce the laws accordingly; one particular law in a developed country by educating the masses may not be an effective technique for under-developed countries where 'cultural dogmas' or 'traditions' are stronger than a 'change'. The case studies can discern the following aspects between developed and developing countries related to their approach to handling visual pollution (Table 4).

In a nutshell, it can be learned that:

- Political will and public sensitisation are the keys behind long-term solutions to visual pollution.
- With legislation, long impacts can be seen.

Table 4. Difference of Approach to Handle Visual Pollution Between Developed and Developing Countries.

Sr. No.	Developed Countries	Developing Countries
1	Awareness and acceptance of the issue	Less realisation of the problem among the public
2	Urge for sustainable solutions	Temporary and ad hoc solutions
3	Coordinated urban development	Fragmented urban development
4	Priority is the beauty of the built environment, public safety, preservation of nature and respect for culture	Priority is revenue generation and political fame
5	Introduction of specific policies, standard operating procedures, acts for different pollutants and land uses.	Absence of dedicated sections of visual pollution in environmental and urban planning laws and policies, and legislation
6	Strict enforcement, carrot-and-stick policies	Weak enforcement of rules and regulations, if available any.
7	Adoption of future approaches and technology for urban settlements to lower down its negative impact	The deficient futuristic vision of urban areas
8	Working on agenda related to the quality of life	Fight for basic necessities of life. Quality of life is far away to ponder about.
9	Effective community participation; they work as working partners and stakeholders.	Less involvement of the community as stakeholders
10	Development control	Shattered and confused development control practices

- Engagement of stakeholders through dialogue and discussion can bring sustainable solutions.
- Urban planning is an umbrella to build harmonised, safe and resilient urban settlements.
- Development control coupled with zoning regulations, land use planning and spatial strategies is the key to avoiding visual pollution.
- Instead of uniform regulations, the introduction of customised and comprehensive plans of all states and localities should be framed to combat the localised nature of visual pollution.

- Better to control and regularise rather than a complete ban, particularly on billboards and hoardings as these are one of the sources of revenue generation specifically for developing economies.
- Awareness, education and training of the inhabitants and officials are the kick-off points regarding the management of visual pollution.
- Cultural change demands repeated and continuous efforts. Therefore, repeated and continuous efforts are required to deal with the spread of visual pollution.

1.4 Conclusion of the Chapter

The urban world embraced multi-issues as the urban population increased since the late nineteenth century and the character of places transformed. The new inductions like city branding, marketing, road network expansion and increased populations made the cities visually distracted by generating more solid waste, slums, outdoor advertisements and utility and infrastructure services. Visual pollution is the resultant product of unmanaged urbanisation by the urban governance that can only be controlled with a combined formula of good urban administration and strict development control flavour with strong political will and comprehensive legislation. Otherwise, the place's natural beauty, individual and personalised identity will be overwhelmed by the man-made chaos.

1.5 Summary of the Chapter

(1) Developed and developing countries are categorised on the GDP and HDI calculation.

(2) Predominantly, the developing countries are located within Asian and African continents with HDI below 0.80, which means these countries' social, urban and economic infrastructures are unstable.

(3) Visual pollution is the agenda of quality of the urban life that is on less priority in developing countries. The consciousness to deal with it is different in developed and developing parts of the world.

(4) The section highlights the examples that are remembered and referred to as the beacon of light for managing visual pollution in different geographic locations.

(5) The Highway Beautification Act 1965, USA, The Clean City Law 2007, Brazil, The Jacksonville Community Study 1983, Florida, the projects of municipalities of Jeddah, Riyad and Al-Hasa 2014–2018, Reforms of Bundaberg Regional Council, Australia and Regulation of outdoor advertisements, Beijing, China are prominent cases that have been captured in this chapter.

(6) Unfortunately, developing countries are struggling with the basic needs of urban dwellers. The identity and visual image of the urban settlement is not their preference.

(7) Although, the voice is being raised by the researchers, environmentalists and social activists against visual pollution; however, it is lacking in legislation and policy steps.

(8) The temporary solutions are introduced by the courts and planning departments of developing countries to support political fame and credibility or to compensate for some human or property loss.

(9) Urban development control is the primary tool to manage visual pollution accompanied by political will, localised set of regulations, and involvement of multi-stakeholders, awareness and advocacy campaigns.

4

VISUAL POLLUTION ASSESSMENT

1. INTRODUCTION

In the context of the research so far discussed, it is clear that the absence of quantitative mechanisms of the visual pollution assessment capturing maximum VPOs (visual pollution objects) is one of the key hindrances for appropriate control of the issue. It is highly pertinent to measure visual pollution and present the information in appropriate spatial visualisation. In recent years, researchers have explored novel ways to address this challenge. The application of modern technologies like mobile-based data collection, geographic information system (GIS)-based visualisations, crowdsourcing, artificial intelligence (AI) and citizen science to identify, assess and map visual pollution is emerging. This chapter presents contemporary developments related to the assessment and quantification of visual pollution, about which knowledge gaps have been identified in chapter two. Further, it presents the discussion on prospects of emerging technologies of AI, crowdsourcing and citizen science with particular frame of visual pollution.

1.1 Assessing the Levels of Visual Pollution

Forming evidence-based policies and management initiatives requires an exact picture and quantum of problems that can come with assessing visual pollution at any desired location and area. Chapter two presented the growth of the visual pollution body of knowledge in six dimensions: conceptualisation, sources, description, application, quantification/assessment and mitigation control. The radar analysis presented in Fig. 10 highlights the skewness of literature available regarding the quantification of the subject. Unlike air, water, noise and land pollution, a major hindrance in assessing visual pollution is the unavailability of a systematic methodology and tool that can record all pollutants of visual pollution collectively, objectively and determine the levels in the form of a standard score.

1.1.1. Visual Pollution Assessment

Visual Pollution Assessment (VPA) is the process of quantifying the level or intensity of visual pollution at any geographical point, node or area of interest by employing a systematic assessment tool. A comprehensive VPA entails recording all understudy VPOs and their characteristics in proportion to the urban form and scale. Such assessment may be completed through manual, semi-automated or fully automated mechanisms such as object-based classification using machine learning.

1.1.2. Visual Pollution Index

In 2020, Wakil introduced the concept of Visual Pollution Index (VPI) as a scale of measuring visual pollution that refers to the levels of visual pollution at any given point measured using Khydija's VPA Tool (Wakil et al., 2019). It is presented as a continuous ratio scale of 0–1, where '0' refers to 'no visual pollution', and '1' refers to the highest level of visual pollution.

1.2 Cases of Visual Pollution Quantification: Street- to City-Scale

The last 10 years have witnessed exceptional growth in geospatial technologies, and IT infrastructure has brought immense computing powers to the palm of every individual. Resultantly, innovative ways have been introduced to apply emerging disruptive technologies in almost every field of life and the subject of visual pollution is no exception. Researchers across the globe have come up with unique ways to apply technological methods to assess, document and map the levels of visual pollution on objects of various scales. Here is a brief indicative summary of the cases where a diverse set of methods have been used to quantify a diverse range of VPOs in a diverse set of urban settings at a diverse scale range. In the context of urban environments, the scale here refers to nodes (a cross point of two or more streets), streets, highways, neighbourhood, city or a district. Authors research published elsewhere (Wakil et al., 2019) documents a comprehensive list of study that contains concepts similar to VPA. Table 5 is adopted from that research. These studies clearly highlight the diversity in VPO coverage, the scale of application and various methods.

1.3 The Challenges of Assessing Visual Pollution

The literature evidence reveals that the subjectivity associated with the visual pollution measurement has been reported repeatedly by the researchers. The quantification of visual pollution at any point has always been challenging because of its subjective nature (Chmielewski, 2020; Planning & State, 2020; Portella, 2007). There does not seem to exist any standard set of guidelines for systematic assessment of visual pollution (Nami, Jahanbakhsh, & Fathalipour, 2016). This further highlights the need for the development of a unified quantitative assessment tool. Although researchers all around the globe have contributed to the assessment of visual pollution, there are a certain set of limitations, which are discussed below:

Micro- vs. Macro-level research: Previous studies on visual pollution can be broadly categorised into two extremes of micro and macro scale. Some previous studies have on a small case study area like a commercial street, public buildings in a residential area, or a neighbourhood with maybe one or two VPOs (billboards, commercial signs). In contrast, other groups of researchers adopted a bigger working scale like a city, with multiple VPOs. Consequently, two (somewhat opposite) strategies with narrow and broad classes of indicators are found in the literature to assess the visual pollution of any area.

Lack of quantification: Since visual pollution is intrinsically sensitive, subjective and a complex type of pollution, no specific tool or instrument is available to measure the scale of visual pollution at any node (Nasar & Hong, 1999; Wakil et al., 2019).

Dependency on subjective variables: Most of the available research has used a mix of subjective and objective indicators where the proportion of subjective indicators has been considerably high. Hence, the results can be potentially influenced either by the respondent or the researchers' interpretation, expertise, interest and prevailing literature concepts. So, they may not be agreed upon or adopted in a different context.

Narrow coverage of VPOs: Visual pollution has a broad area of knowledge of built and aesthetic environments comprising multiple objects. However, most previous work has focused on measuring visual pollution employing single VPOs, which has resulted in single-object assessment approaches.

Table 5. List of Studies Containing Components Like Visual Pollution Assessment.

Sr.	Study	Scale	VPOs Coverage	Methods Employed
1	Visual Preferences in Urban Signscapes (Nasar & Hong, 1999)	City	Single VPO; Signs	Colour Photograph
2	Evaluating Commercial Signs in Historic Streetscapes: the Effects of the Control of Advertising and Signage on User's Sense of Environmental Quality (Portella, 2007)	Street	Single VPO; Commercial signs, (outdoor advertisements)	Opinion survey
3	Evaluation of visual pollution in urban squares, using SWOT, AHP and QSPM techniques (Allahyari, Nasehi, Salehi, & Zebardast, 2017)	Neighbourhood	Multiple VPOs; Outdoor advertisements, Garbage, Congestion, Graffiti, Absence of green spaces, Building heights	AHP, QSPM and SWOT (Strengths, Weaknesses, Opportunities, Threats)
4	Measuring visual pollution by outdoor advertisements in an urban street using inter-visibility analysis and public surveys (Chmielewski, Lee, Tompalski, Chmielewski, & Wężyk, 2016)	Street	Single VPO; Outdoor advertisements	Inter-visibility analysis Public survey
5	Introduction to a quantitative method for assessment of visual impacts of Tehran Towers (Karimipour, Mojtahedi, & Dehkordi, 2015)	City	Single VPO; Cell Towers	Quantitative, Visibility Analysis using GIS

Table 5. (*Continued*)

Sr.	Study	Scale	VPOs Coverage	Methods Employed
6	Visual pollution can have a deep degrading effect on urban and suburban community: a study in a few places of Bengal, India, with special reference to unorganised billboards (Jana, 2015)	District	Single VPO; Billboards	Visual comparisons
7	Citizen science and WebGIS for outdoor advertisement visual pollution assessment (Chmielewski, Samulowska, Lupa, Lee, & Zagajewski, 2018)	City	Single VPO; Outdoor advertisements	Opinion survey Visual pollution score Spatial mapping
8	Urban Environmental Graphics: Impact, Problems and Visual Pollution of Signs and Billboards in Nigerian Cities (Bankole, 2013)	City	Single VPO; Billboards	Colour photos
9	Examining Impact of Visual Pollution on City Environment: Case Study of Pune, India (Gokhale, 2011)	City	Multiple VPOs; Hoardings, Billboards, Dustbins, Utility Wires, Light Poles, Parking	Public Opinion Survey
10	Free Standing Billboards in a Road Landscape: Their Visual Impact and Its Regulation Possibilities (Lithuanian Case) (Jurate Kamicaityte-Virbasiene, 2013)	State road	Single VPO; Free Standing Boards (FSB)	Orthophoto Maps, Field Survey

Table 5. *(Continued)*

Sr.	Study	Scale	VPOs Coverage	Methods Employed
11	Visual pollution and statistical determination in some of Karrada district main streets, Baghdad (Atta, 2013)	Street	Multiple VPOs; Garbage, electric wires, military weapons, demolished buildings, excavation works and rubbles, billboards etc.	Public Opinion, Statistical Analysis
12	Regulating outdoor advertisement boards; employing spatial decision support system to control urban visual pollution (Wakil, Hussnain, Naeem, & Tahir, 2016)	Primary road	Single VPO; outdoor advertisement	GIS

The absence of structured tools: Unlike other types of pollution, structured instruments and tools are not available to measure visual pollution. Moreover, in some cases, a combination of different kinds of methodologies like public surveys (Nami et al., 2016; Ogunbodede & Sunmola, 2014), inter-visibility analysis, triangulation method, focus group discussions, photo comparisons (A. A. Portella, 2007; Voronych, 2013), visual comparisons and experimentation etc. (Jana & De, 2015b) has been explored to assess a few (or often a single) VPO, which again indicates the need of a composite VPA tool. In most cases, the statistical validity or reliability of the methods has not been assessed and discussed thoroughly.

1.4 Tools and Methods for Visual Pollution Assessment

Researchers worldwide are working on the different aspects of visual pollution by employing different methodologies and tools. These varying techniques include colour photographs, opinion surveys, public surveys, Analytical Hierarchy Process (AHP), Quantitative Strategic Planning Matrix (QSPM), Strength Weaknesses Opportunities and Threats (SWOT), inter-visibility and quantitative analysis using GIS, spatial mapping, orthophoto maps, visual comparisons, visual pollution scoring, statistical analysis, landscape metrics analysis and 3D Isovist and Voxel Approach.

1.5 The Recent Development on Visual Pollution Assessment

In 2019, the authors introduced a systematic approach to quantify visual pollution. Through our earlier research work published elsewhere (Wakil et al., 2019), we present an android-based methodology called VPA tool, which provides a structured mechanism for the quantification of visual pollution at any given location by assessing the presence and characteristics of various VPOs on a scale of 0–1.

The development of this VPA tool resulted in the output of a PhD research where the comprehensive list of VPOs has been formed through literature review, photo competitions of university students and a public opinion survey. The VPOs were classified into 10 major groups by the sectoral specialist who ranked the VPOs based on their contribution to visual pollution levels. The subjectivity of the indicators has been handled by the involvement of expert panels through AHP. For each VPO, a set of rubric values have been defined, and finally, the VPA tool is formed as a combination of an android-supported VPA scorecard and VPA calculator, which is commonly known as Khydija's VPA tool (Fig. 12). The VPA scorecard records the presence of available VPOs and their characteristics out of 39 VPOs, covered in the tool (Fig. 13).

VPA calculator counts the total VPA score by using VPA scorecard information with the following formula/equation:

Max attainable score: a = Sum of highest scores under each characteristic

The score attained: b = Sum of VPO specific score attained

VPO Specific Ratio Score: $c = a \div b$

Weighted score: $d = c \times$ VPO type weight

Overall weighted score: $e = d \times$ VPO group weight

1.5.1. Main Characteristics of Khydija's VPA Tool

The key characteristics of the tool are:

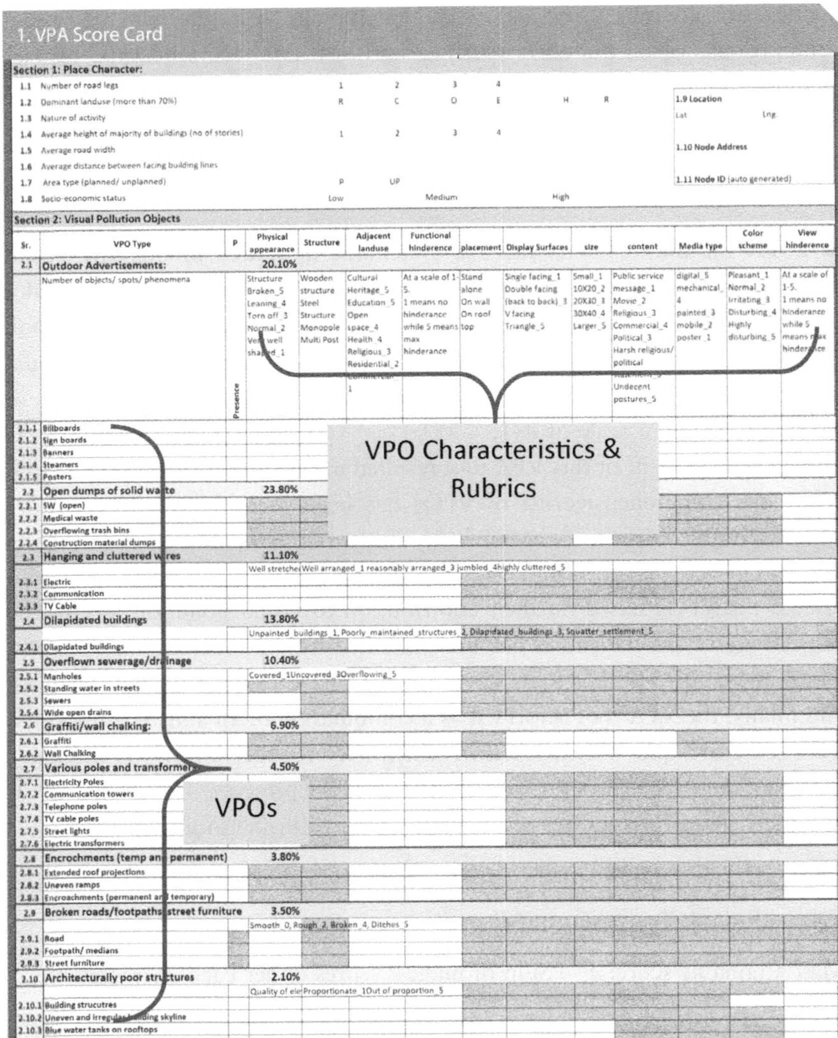

Fig. 12. Snapshot of the Visual Pollution Assessment Tool.

- It offers a structured mechanism for quantifying visual pollution, which is available in both paper-based and android-based tools.
- It has a broader coverage of 39 VPOs that increase its application in wider parts of the world, especially the developing countries.
- It records the presence and characteristics of VPOs to assess/quantify visual pollution.

2. VPA Calculator

Sr.	VPO Type	Weight	Max attainable score	Score attained	VPO Specific Ratio Score	Weighted score	Overall weighted score
Section 1: Place Character:							
Node ID							
Section 2: Visual Pollution Objects							
2.1	**Outdoor Advertisements:**	**20.10%**					
	Number of objects/ spots/ phenomena		Sum of highest scores under each characteristics	Sum of VPO specific score attained	Ratio of attained score to max score	Ratio score multiplied by VPO Type weight	Weighted score multiplied by VPO Group weight
2.1.1	Billboards	55%	60				
2.1.2	Sign boards	20%	60				
2.1.3	Banners	10%	60				
2.1.4	Steamers	5%	60				
2.1.5	Posters	10%	60				
2.2	**Open dumps of solid waste**	**23.80%**					
2.2.1	SW (open)	35%	25				
2.2.2	Medical waste	20%	25				
2.2.3	Overflowing trash bins	25%	25				
2.2.4	Construction material dumps	20%	25				
2.3	**Hanging and cluttered wires**	**11.10%**					
2.3.1	Electric	70%	30				
2.3.2	Communication	20%	30				
2.3.3	TV Cable	10%	30				
2.4	**Dilapidated buildings**	**13.80%**					
2.4.1	Dilapidated buildings	100%	20				
2.5	**Overflown sewerage/drainage**	**10.40%**					
2.5.1	Manholes	25%	25				
2.5.2	Standing water in streets	15%	15				
2.5.3	Sewers	25%	20				
2.5.4	Wide open drains	35%	20				
2.6	**Graffiti/wall chalking:**	**6.90%**					
2.6.1	Graffiti	50%	40				
2.6.2	Wall Chalking	50%	40				
2.7	**Various poles and transformers**	**4.50%**					
2.7.1	Electricity Poles	25%	30				
2.7.2	Communication towers	10%	30				
2.7.3	Telephone poles	20%	30				
2.7.4	TV cable poles	5%	30				
2.7.5	Street lights	20%	30				
2.7.6	Electric transformers	20%	30				
2.8	**Encrochments (temp and permanent)**	**3.80%**					
2.8.1	Extended roof projections	35%	20				
2.8.2	Uneven ramps	50%	20				
2.8.3	Encroachments (permanent and temporary)	15%	20				
2.9	**Broken roads/footpaths/street furniture**	**3.50%**					
2.9.1	Road	40%	15				
2.9.2	Footpath/ medians	30%	15				
2.9.3	Street furniture	30%	15				
2.10	**Architecturally poor structures**	**2.10%**					
2.10.1	Building strucutres	40%	35				
2.10.2	Uneven and irregular building skyline	30%	30				
2.10.3	Blue water tanks on rooftops	15%	45				
2.10.4	Dish receivers and broken antenas	15%	45				
	Cumulative Visual Pollution Score (Sum of Overall weighted score of all VPO types)						

$$VPI = \sum_{n=1}^{n=VPOs} \left\{ \left(\frac{d}{c} \right) \times b \times a \right\}$$

a = VPO group weight

b = VPO weight

c = Max attainable score i.e. sum of highest scores under each characteristic

d = Attained score i.e. sum of VPO specific score attained

Available at:
https://urbanvisualpollution.wordpress.com/visual-pollution-measurement-vpm/

Fig. 13. Visual Pollution Assessment Calculator.

- All the collected data are geotagged and can generate visualisation for the spatial spread of visual pollution.
- From the analysis, it is possible to ascertain the primary contributor VPOs per location. Resultantly, targeted action can be taken to fix the visual pollution problem.
- Users can record photos for each VPO, which can be used to verify by the visual pollution data analyst.
- Since the data are collected at point level, it is possible to aggregate results at any upper tier such as a street, neighbourhood, city or district. It can be applied irrespective of the geographical scale of the case study.

1.6 Application of the Tool

The Khydija's VPA tool can help urban planners and government stakeholders to understand the prevalence of visual pollution better, assess its spatial spread, identify primary contributing VPOs per location, take any corrective actions and (most importantly) inform policy decisions in a robust, quantifiable and evidence-driven manner.

The tool has been implemented in the urban area of Rawalpindi, a city in one of the developing countries of South Asia, Pakistan. It is the fourth-largest metropolitan city in Pakistan with a population of more than 3 million (CityPopulation, 2017; Government of Pakistan, 2018). The city is situated at the 33°36′N 73°02′E and serves as the divisional capital to the Rawalpindi division within Punjab, the most populated province of Pakistan (Fig. 14). The city population has increased by 48.8% since the previous census of 1998, with an annual growth rate of 2.11% per year. The details on the city limits and administrative bodies have been presented in (Fig. 15). It presents that the case study area is a mix of different socioeconomic groups residing in planned and unplanned settlements located and governed by different public and private controlling agencies including Rawalpindi Development Authority, City District Government, and Defence Housing Authority and Cantonment.

1.7 Spatial Sampling for the Visual Pollution Assessment

It is commonly accepted and believed that a simple random sample can be less effective when the target population contains spatial autocorrelation and stratified heterogeneity. The spatial sampling approach handles these characteristics of the data (Wang, Stein, Gao, & Ge, 2012; Wang, 2018). The author's PhD research work suggests that spatial sampling is desirable and

Fig. 14. Location Map of the Study Area in the Regional Context.

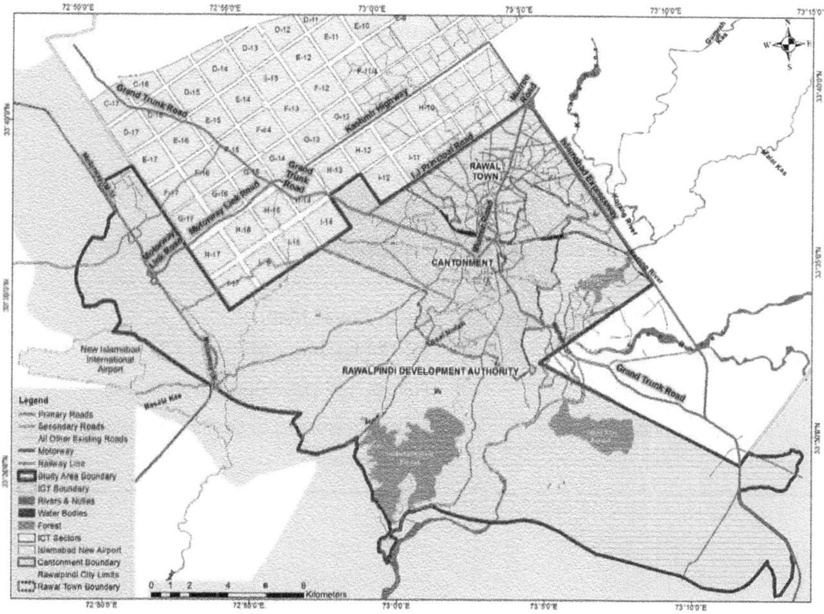

Fig. 15. Study Area Limits.

preferred in the case of VPA as well. In this research, the 'population' refers to the total nodes generated as the intersection points of the single-line road network data obtained from the study area's Open Street Map (OSM). Spatial sampling has been used to select a sample from spatially distributed nodes as per the strata requirements in the previous section and then infer parameters of the target nodes, such as the mean and values at unsampled sites. The following process has been used to obtain a spatially distributed sample.

1.7.1. Acquisition of Data Layers Representing Strata

The urban Rawalpindi base map has been prepared using different data layers as shown in Table 6. The road network data obtained from OSM has been used to calculate the total number of nodes (intersections of roads). These nodes serve as the total universe of locations where urban visual pollution can be accessed. It has been generated using the 'Creating a point feature at an intersection' tool in ArcGIS (ESRI, 2020). This process resulted in the generation of 162,619 nodes across the study area. Using the confidence level of 95% and a confidence interval of 5, a sample size of around 383 nodes is required for VPA in Rawalpindi. On the other hand, based on the process

Table 6. Spatial Data Layers and Sources.

Sr. No	Data Layer	Source
1	City limits	Rawalpindi Development Authority
2	Union councils boundaries	City District Government
3	Cantonment limits	City Pulse Pvt. Ltd.
4	Municipal Corporation limits (Rawal town)	Rawalpindi Development Authority
5	Road network (single line)	Open Street Map
6	Building blocks	City Pulse Pvt. Ltd
7	Land use	Rawalpindi Development Authority
8	Socioeconomic status	Pakistan Bureau of Statistics, Pakistan Census Organisation

adapted from Hogg, followed by the sample size determinisation, 160 points were needed (Hogg, 2016; Hogg, Tanis, & Zimmerman, 2010). In addition to the base map, the spatial data on the strata layers have been collected from various public and private resources.

1.7.2. Demarcation of Clusters Representing Sampling Strata

The city area has been divided into larger clusters representing stratum each for an urban phenomenon. Each urban block has been assigned with spatial attributes including dominant land use (residential, commercial, others), controlling authority (Rawalpindi Development Authority, Tehsil Municipal Administration, Cantonment, Defence Housing Authority etc.), area type (planned, unplanned) and socioeconomic class. Overlay techniques have been used in the ArcGIS tool to generate the strata. A series of screenshots representing each data layer has been presented in the following sections (Figs. 16–20).

1.7.3. Finalisation of the Spatial Sample

The sample size of 303 locations has been identified on the map to cover the statistically significant numbers of nodes from each stratum, using the spatial sampling approach (Fig. 21).

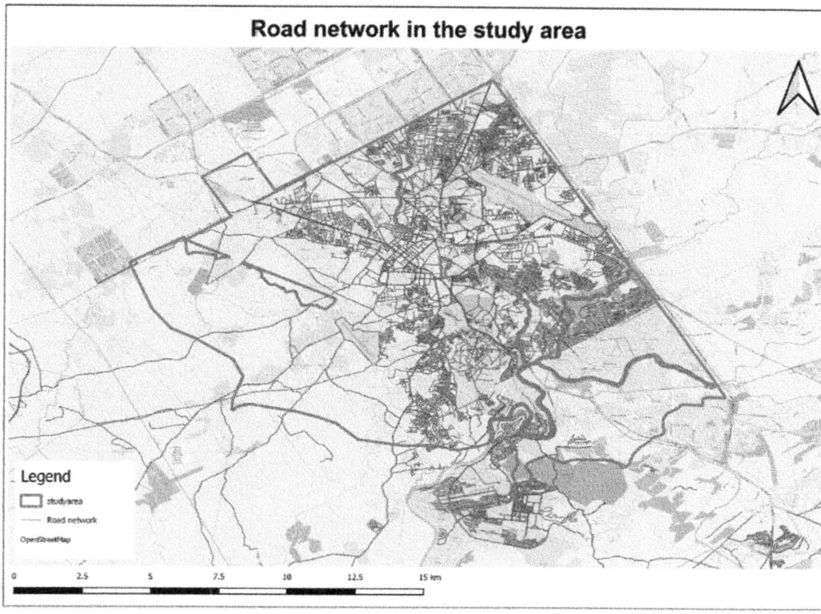

Fig. 16. Single-Line Road Network in the Study Area.

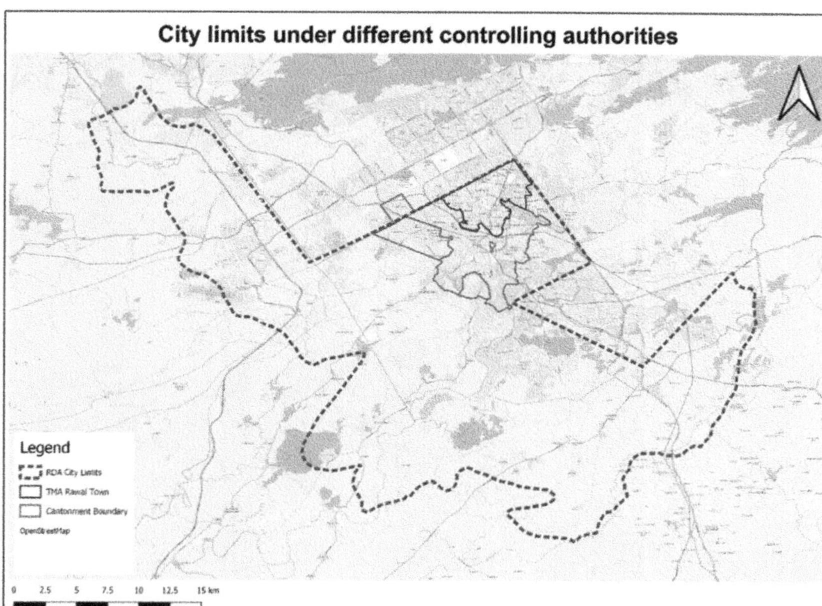

Fig. 17. City Limits Under Different Controlling Authorities.

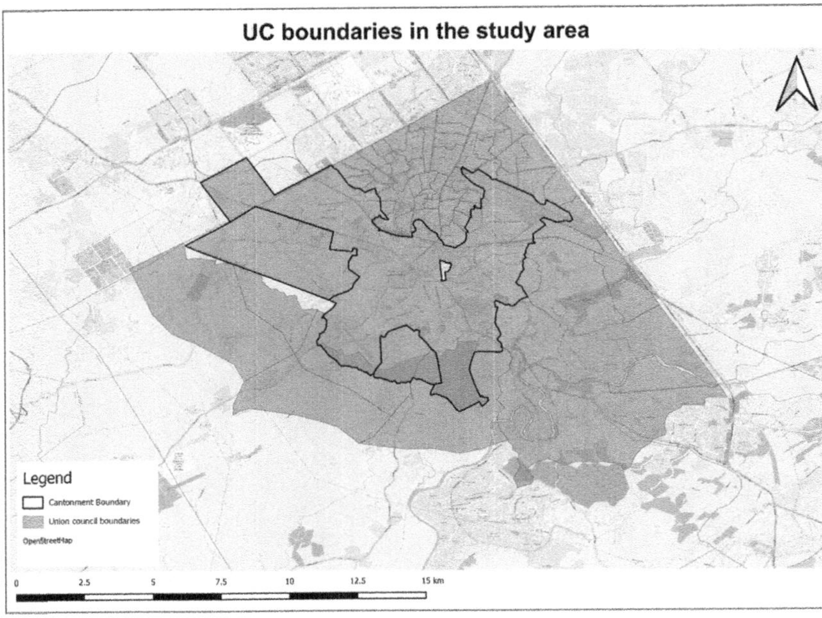

Fig. 18. Union Council Boundaries in the Study Area.

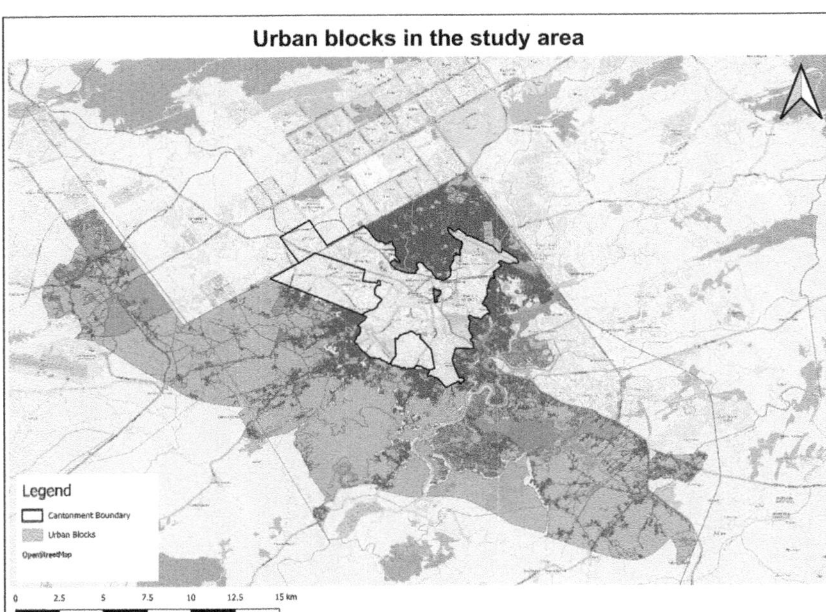

Fig. 19. Map Showing the Distribution of the Study Area into Urban Blocks.

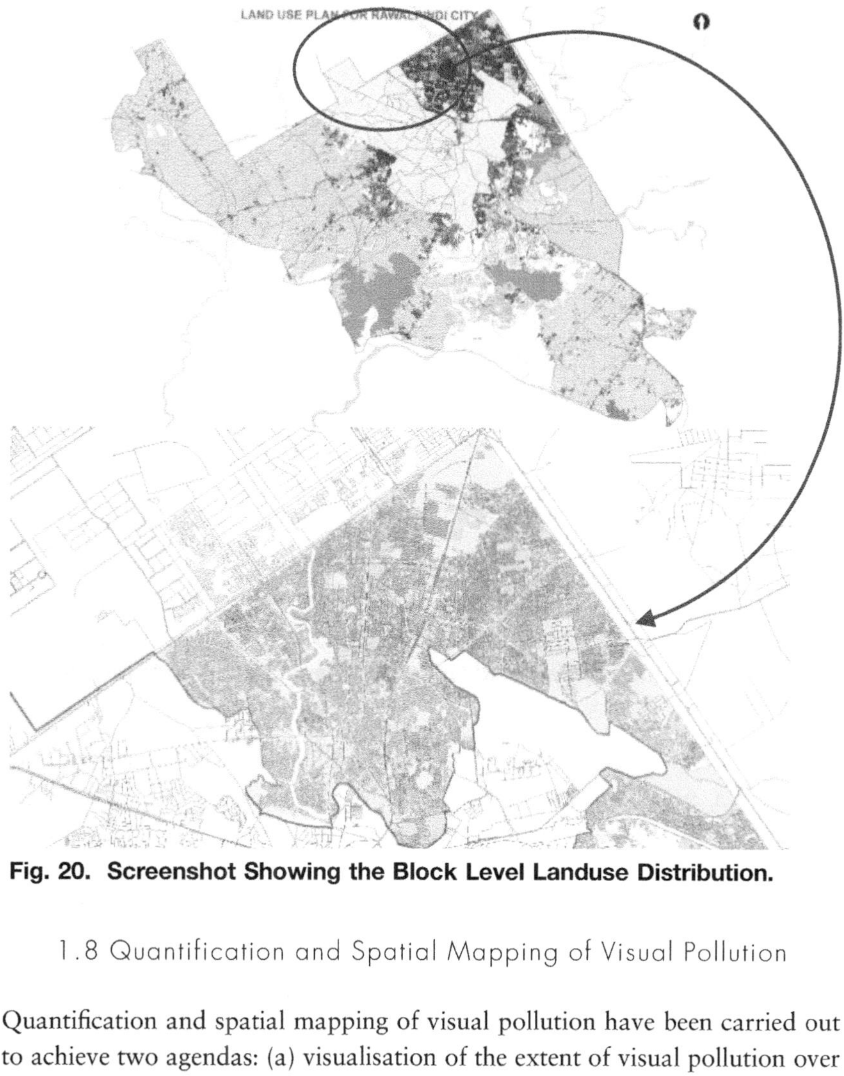

Fig. 20. Screenshot Showing the Block Level Landuse Distribution.

1.8 Quantification and Spatial Mapping of Visual Pollution

Quantification and spatial mapping of visual pollution have been carried out to achieve two agendas: (a) visualisation of the extent of visual pollution over space and (b) understanding the relationship between the prevalence of visual pollution and other urban phenomena, including land-use type, population density, socioeconomic class and controlling authority. Following, the methodological process has been used for this purpose:

1.8.1. Field Survey

During the field survey execution, a team of seven enumerators (five men and two women) was hired to conduct the VPA. Each enumerator was assigned a

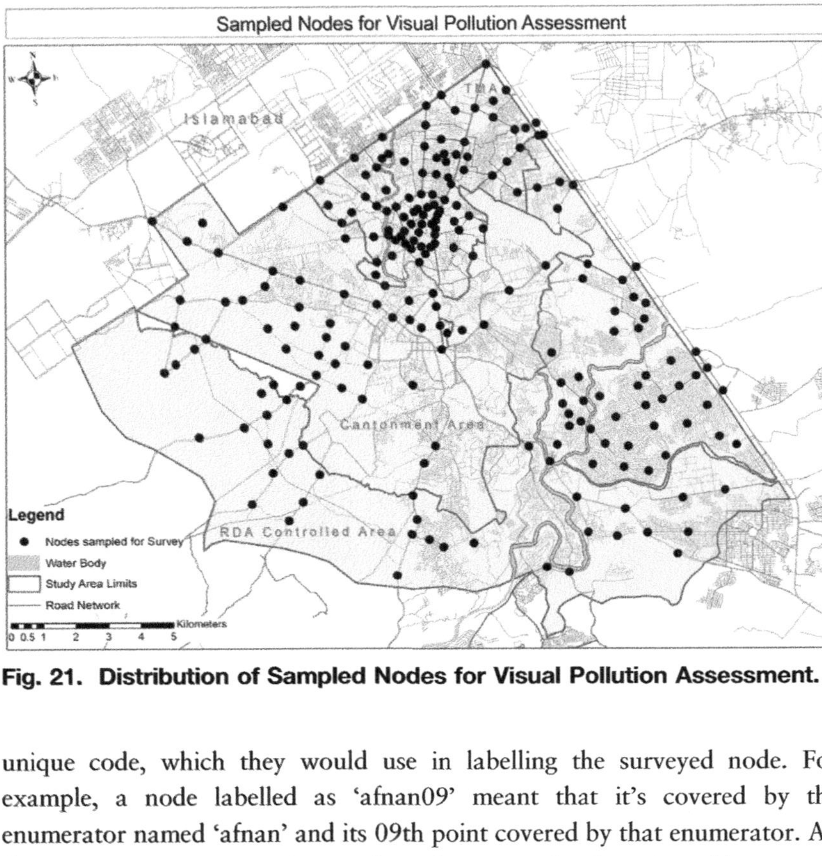

Fig. 21. Distribution of Sampled Nodes for Visual Pollution Assessment.

unique code, which they would use in labelling the surveyed node. For example, a node labelled as 'afnan09' meant that it's covered by the enumerator named 'afnan' and its 09th point covered by that enumerator. All the enumerators were graduate urban planners trained on the subject assessment and the tool employing a three-hour training workshop. The training workshop was held on 19 April 2018, where each enumerator was asked to fill test forms for any five nodes around the training venue to ensure that they were well fluent in using the tool. Every enumerator was equipped with an android device having Open Data Kit (ODK) Collect (ODK, 2020) installed along with the VPA tool. In addition to the mobile app, the enumerators were provided with the paper-based VPA tool and a map of the sampled locations (Fig. 21).

The sampled locations were provided in KML (Keyhole Markup Language) format to reach an assigned location using their cell phones. The map of sampled nodes was divided into seven clusters with the manual interpretation of the city so that each enumerator may follow a set of the connected road to reach the sampled location. Each enumerator was provided with a personalised authority letter on behalf of the National University of Sciences and Technology to help them move across the study area to handle the prevailing security situation. The survey spanned on a five working-day schedule. During the survey days, the author acted as a control centre with on-call contact with all the enumerators. The data collected was sent to the specified server in real time, monitored by the author for quality control. For disseminating two-way communication, a WhatsApp group named Visual Pollution Survey was used. The use of a chat group helped disseminate any instructions from the control station to the enumerators and receive any feedback from the field teams. During the VPA tool preparation and validation, it was found that the average time for VPO capturing at one node has been 10–15 minutes. Based on the initial analysis of the VPA tool application at a single node, every enumerator was tasked to cover approximately 10 nodes per day.

The enumerators were tasked to visit the designated node and stand with their face towards the north and start taking a 360-degree view of the site in a clockwise direction while noting down the details of each VPO. They were advised to ensure that in case of inaccessibility of location, the new points should be evenly distributed and MUST NOT be concentrated along main roads only. Furthermore, the enumerators were instructed regarding the spatial coverage that if they face any challenge in accessing a particular location, they may choose a nearby location considering the same strata aspects.

1.8.2. Data Analysis

The data collected from the VPA tool have been analysed to generate a VPI at each surveyed point. Following, statistical processes have been applied to the data:

(1) Correlation

- Visual pollution level vs. area type
- Visual pollution level vs. the controlling agency
- Visual pollution level vs. socioeconomic status of the area
- Visual pollution level vs. land uses

(2) Descriptive statistics

(3) Inferential statistics

1.9 Visual Pollution Distribution Across Urban Phenomena

The data on VPO existence has been converted to the VPI based on the calculation methods explained in the methods section. Based on the characteristics of rubrics of each VPO, and respective VPO group weight, a VPO score has been calculated for each VPO. Furthermore, based on the cumulative VPO score, VPI has been calculated. Finally, the average cumulative score of VPOs and VPI has been determined for every node. Table 7 presents the average cumulative score of VPOs and VPI across each urban phenomena under study. The most important part of the table is the average VPI value, reflecting the pattern. The data reveal that among the various land uses the level of visual pollution is the highest at commercial nodes (0.5124), followed by public buildings at 0.4600 and educational buildings at 0.4109. The level of VPI at the park and open spaces and vacant land is measured at 0.3610 and 0.3129, respectively. The lowest value of VPI is measured at the residential land use, i.e. 0.2831 (Fig. 22).

On the other hand, in the case of the planned and unplanned areas, the average VPI value stands at 0.2858 and 0.3912, clearly showing a higher level of VPI in the unplanned areas. Similarly, in the case of controlling authorities, the VPI is highest in Tehsil Municipal Administration (TMA) controlled areas (0.4385), followed by RDA with 0.3843, while cantonments/DHA (Defence Housing Authority) has the lowest value of 0.3298. A very similar and crisp pattern is visible among areas of different economic status. The lowest VPI average is measured at the high economic status areas at 0.2713 and medium economic status areas at the middle with 0.3617. In contrast, the highest level of VPI is found in the low economic status area with a value of 0.4540.

1.9.1. Descriptive Statistics

Though the various correlation tests do not reflect a significant correlation between VPI and area type or between VPI and land uses, it was important to look deeper at the other patterns. The descriptive analysis, box and whisker plots, and histograms have been prepared to understand the deeper patterns.

Descriptive statistics are brief descriptive coefficients that summarise a given dataset, representing the entire or a sample of a population. Descriptive statistics are broken down into measures of central tendency and measures of variability. Measures of central tendency include the mean, median and mode, while measures of variability include standard deviation, variance, minimum and maximum variables, and kurtosis and skewness.

Table 7. Average of the Cumulative Score of VPOs and VPI Across Various Urban Phenomena.

Urban Phenomena	Categories	Average of the Cumulative Score of VPOs and VPI										
		Outdoor Advertisements	Solid Waste	Hanging Wires	Dilapidated Buildings	Sewerage/ Drainage	Graffiti	Poles/ Transformers	Encroachments	Broken Roads	Architecturally Poor Structures	Average of VPI
Land use	Commercial	0.1838	0.0513	0.0552	0.0212	0.0194	0.0037	0.0437	0.0167	0.0080	0.0029	0.5124
	Educational	0.1613	0.0444	0.0729	0.0148	0.0305	0.0026	0.0408	0.0153	0.0115	0.0003	0.4109
	Vacant land	0.2008	0.0557	0.0508	0.0152	0.0171	0.0063	0.0450	0.0158	0.0089	0.0031	0.3921
	Parks/Open spaces	0.1146	0.0502	0.0466	0.0098	0.0404	0.0077	0.0420	0.0138	0.0135	0.0017	0.3610
	Public buildings	0.1510	0.0565	0.0505	0.0145	0.0170	0.0024	0.0374	0.0153	0.0060	0.0016	0.4600
	Residential	0.1342	0.0641	0.0419	0.0096	0.0163	0.0039	0.0384	0.0143	0.0057	0.0010	0.2831
Area type	Planned	0.1810	0.0500	0.0505	0.0141	0.0193	0.0034	0.0429	0.0154	0.0075	0.0017	0.2858
	Unplanned	0.1650	0.0559	0.0554	0.0213	0.0202	0.0040	0.0419	0.0166	0.0081	0.0028	0.3912
Controlling authority	Cantoment/ DHA	0.1941	0.0539	0.0561	0.0237	0.0242	0.0036	0.0450	0.0163	0.0095	0.0033	0.3298
	RDA	0.1688	0.0590	0.0504	0.0163	0.0178	0.0044	0.0421	0.0153	0.0082	0.0020	0.3843
	TMA	0.1689	0.0486	0.0543	0.0169	0.0193	0.0033	0.0417	0.0166	0.0067	0.0022	0.4385
Economic status	High	0.1656	0.0395	0.0451	0.0091	0.0153	0.0034	0.0421	0.0148	0.0048	0.0015	0.2713
	Medium	0.1671	0.0679	0.0565	0.0213	0.0246	0.0053	0.0416	0.0169	0.0098	0.0029	0.4540
	Low	0.1762	0.0528	0.0553	0.0205	0.0196	0.0033	0.0427	0.0162	0.0082	0.0025	0.3617

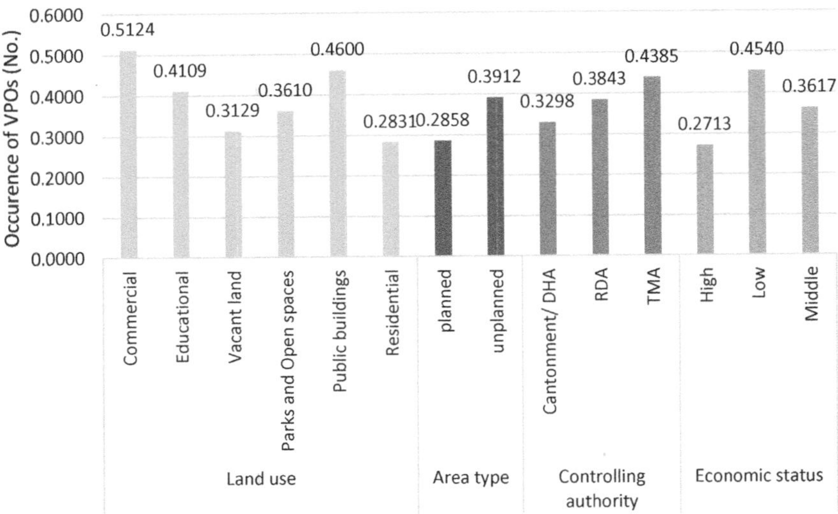

Fig. 22. Average Visual Pollution Index for Various Categories of Urban Phenomena.

The analysis reveals that both the lowest and the highest values of VPI for unplanned areas (0.134, 0.812) are higher than that of the planned area (0.505, 0.5979). Similarly, the median value of unplanned areas (0.351) is also higher than that of planned areas (0.270). Finally, it has been noted that the upper quartile of the VPI in the unplanned area is more expanded than that of the planned area, reflecting the presence of higher visual pollution in unplanned communities (Figs. 22 and 23).

In the case of the area's economic status, the lowest value of VPI in the low economic status area (0.175) is quite higher than those from medium (0.0351) or high economic status areas. (0.0589). Similarly, the upper value in low economic status is noted at 0.746, followed by 0.6665 for the medium economic status and 0.4227 for the areas of high economic status as shown in Table 8. Box and Whisker plot of the statistics in Fig. 24 shows that the distribution is most widely spread in the medium economic status area and the shortest in the high economic status areas. Furthermore, the histogram chart reflects that in the case of high economic status area, most of the nodes fall in the range of 0.3–0.4, while in the case of medium economic status, larger group of the nodes have a VPI value from 0.4 to 0.5. On the other hand, in the case of low economic status, the VPI values are from 0.4 to 0.6, while some of the nodes have values as higher as 0.8 as well (Figs. 25 and 26).

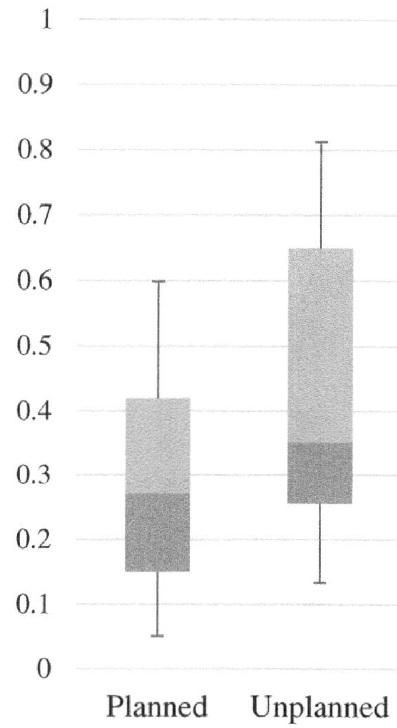

Fig. 23. Visual Pollution Distribution with Respect to Area Type.

From the perspective of controlling authority, the lowest of the minimum VPI value is found at the RDA (0.1589), followed by cantonment/DHA (0.1750), while the highest of the min value is that of TMA (0.2927). It refers to the fact that even the lowest level of visual pollution in the TMA area would be higher than other controlling authority areas (Fig. 27).

Table 8. Descriptive Statistics of VPI for Area Type, Economic Status and Controlling Authority.

	Area Type		Economic Status			Controlling Authority		
	Planned	Unplanned	Low	Medium	High	RDA	TMA	Cantonment
Min	0.0505	0.134	0.175	0.0351	0.0589	0.1589	0.2927	0.1750
Q1	0.150	0.255	0.327	0.2732	0.2099	0.3143	0.3728	0.2229
Median	0.270	0.351	0.445	0.3663	0.2647	0.3838	0.4277	0.3039
Q3	0.4198	0.65	0.592	0.4616	0.3104	0.5341	0.5196	0.3717
Max	0.5979	0.812	0.746	0.6665	0.4227	0.7058	0.7665	0.5372

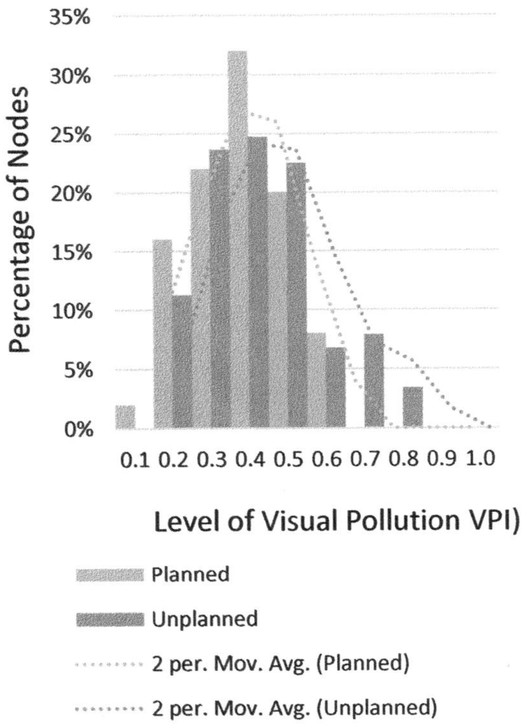

Fig. 24. Histogram of Visual Pollution Distribution vs. Area Type.

The same is true in the case of the median value, which is the highest in the case of TMA (0.4277), followed by RDA with 0.3838 and cantonment at 0.3039. The histogram provides a further insight that in the case of cantonment/DHA, a larger number of nodes have a VPI value of 0.3–0.4, while in the case of TMA, most of the nodes contain a VPI value from 0.3 to 0.5. On the other hand, in the case of RDA, most of the nodes have a VPI value from 0.3 to 0.5, with a higher number of nodes at 0.3 (see Figs. 26 and 27).

From the perspective of land uses, the descriptive statistics of Table 9 presents many interesting patterns. The analysis reveals that for residential land use the minimum VPI value has been recorded at 0.1505, for commercial 0.2094, for public buildings 0.1351, for education 0.2986, while for parks and open spaces, the minimum VPI is 0.3059. Similarly, the median values are calculated to be 0.3647, 0.4613, 0.6236, 0.5236, 0.5236, respectively. It is evident that the median value is highest for commercial (0.5124), followed by 0.4600 for VPI in public buildings, 0.4109 for educational and 0.3921 for open spaces. At the same time, the smallest VPI median has been noted in the residential areas (Figs. 27–29).

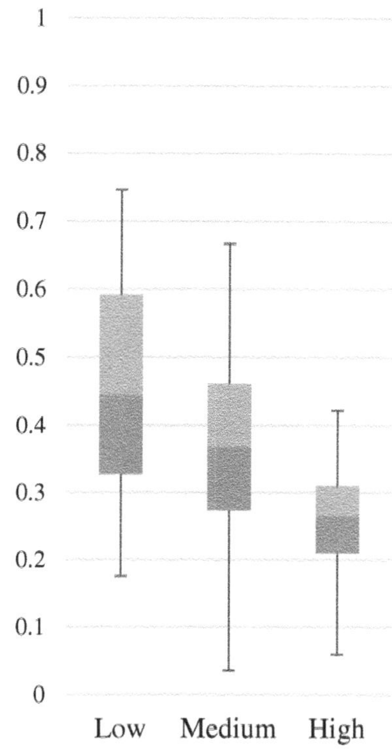

Fig. 25. Visual Pollution Distribution Across the Economic Status of the Area.

1.10 Mapping of the Visual Pollution in the Study Area

This section showcases the research's spatial outputs that demonstrate the variety of analyses extracted using the data (Table 10) collected through the VPA tool. The different map demonstrations include:

- Presence of VPOs using scaled symbols: It is also called proportional symbols map, which means the bigger the size of the dot, the more VPOs present at that particular location and on average 8.7 VPOs exist at each node. Further, the highest concentration goes up to 16 at Saddar Bazar junction and Railway station junction (Fig. 30)
- Node level visual pollution score: This output presents the concentration and visual pollution score at all nodes. Again, dot-scaled map has been used to show the results, and it is observed that the VPI is particularly high along the major roads (Fig. 31).

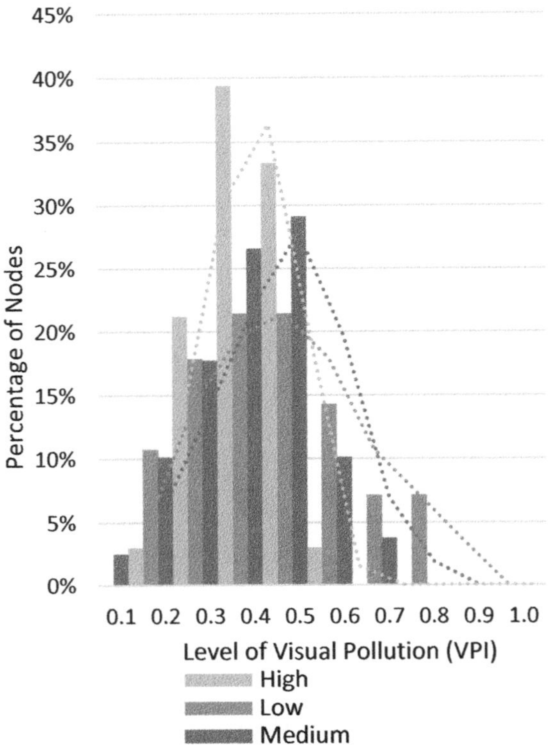

Fig. 26. Histogram of Visual Pollution Distribution vs. Economic Status.

- Average VPI scores generalised at union council level: The choropleth map shows the union council level visual pollution concentration on a scale of 0–1. The average of the VPI score of all the nodes falling in each union council has been used. The green means areas with zero or low visual pollution and red stands for the high or maximum visual pollution. Khanna pul, PWD, Saddar and Satellite towns are some of the most polluted union councils on the said scale (Fig. 32).

- Union council VPI scores viz-a-viz nodal VPI score: This graphical output combines two outputs, including visual pollution spread at union council and nodal levels. It helps comprehend the point-wise individual and union council collective proliferation of visual pollution (Fig. 33).

- Heat mapping of the case study area: This research demonstration reflects the density of visual pollution that is particularly facilitated in better administration and realisation of the issues by presenting its intensity. It is also found that PWD road, Murree Road and Pirwadhai road have high linear concentrations of visual pollution (Figs. 34 and 35).

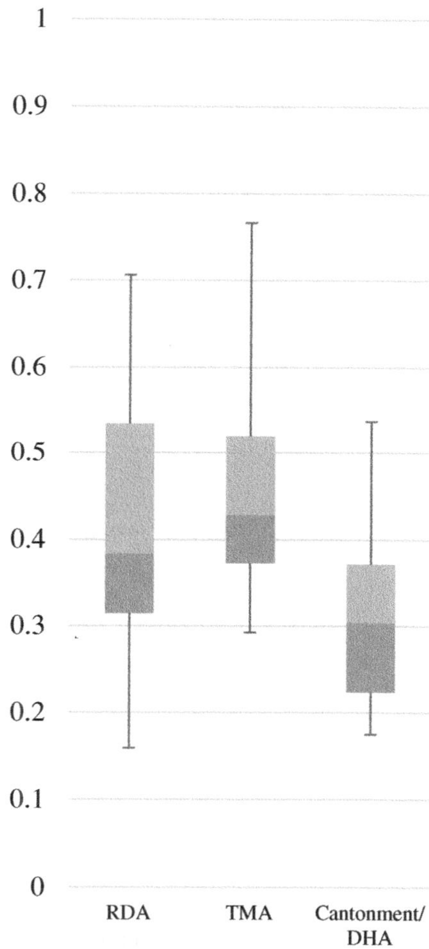

Fig. 27. Visual Pollution Distribution Across Controlling Authorities.

Table 9. Descriptive Statistics of VPI for Various Land Uses.

	Residential	Commercial	Public Building	Education	Open Space
Min	0.1505	0.2986	0.1351	0.2094	0.3059
Q1	0.2125	0.3842	0.3515	0.3919	0.3235
Median	0.2831	0.5124	0.4600	0.4109	0.3921
Q3	0.5167	0.5689	0.6689	0.5545	0.5900
Max	0.7189	0.6747	0.7747	0.7665	0.6307

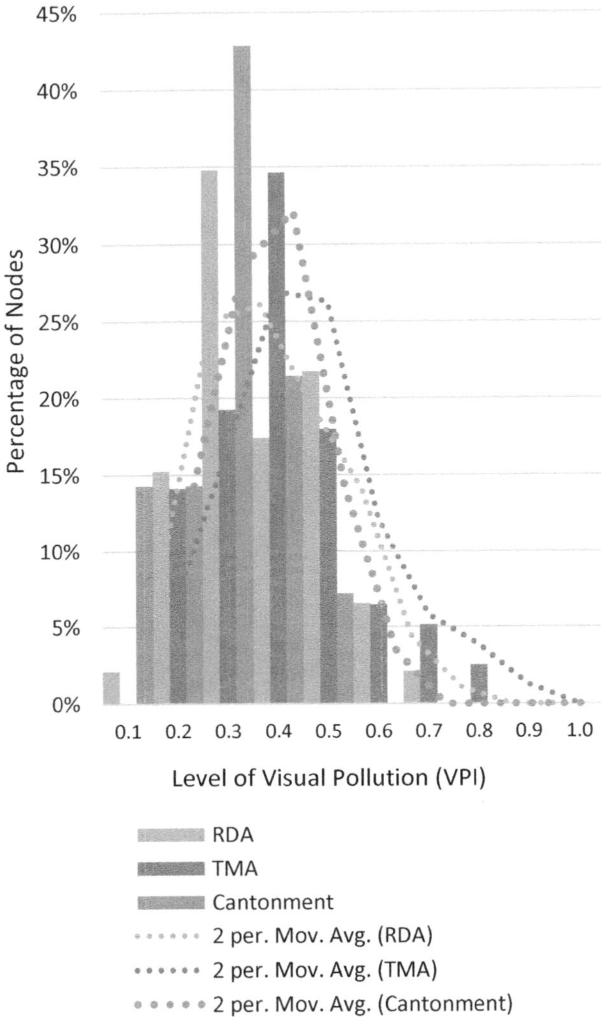

Fig. 28. Histogram of Visual Pollution Distribution vs. Controlling Authorities.

1.11 The Engagement of Modern Technologies in the Subject of Visual Pollution

These are the collection of computer-based, information technology-driven machine brains hired to engage public participation to collect data with better understanding, precision and quality. Artificial intelligence, machine learning, crowdsourcing and citizen science may potentially apply to visual pollution.

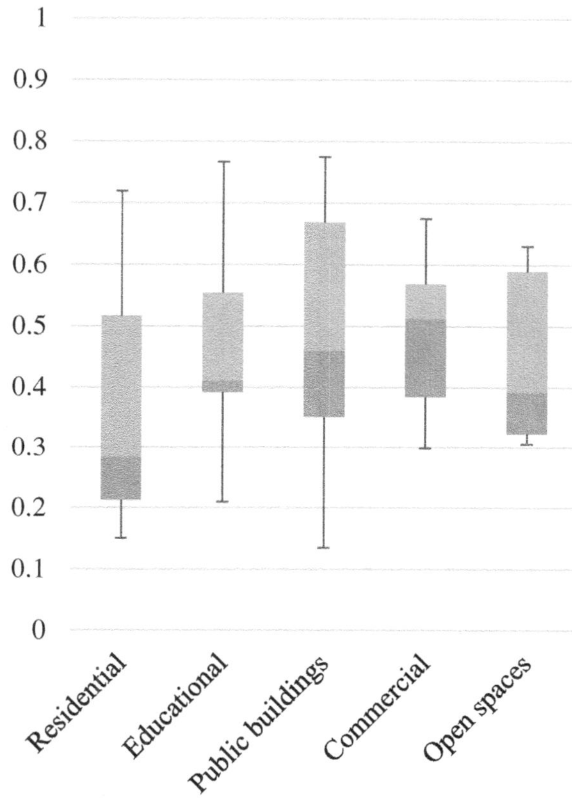

Fig. 29. Visual Pollution Distribution with Respect to the Land Uses.

The quantification of VPO in any area is essential information for management or regulatory bodies to analyse and accordingly take regulatory and precautionary actions. Earlier in the sections, it was highlighted how to gather the right amount of data to visualise the results and take effective actions. The present section takes another approach and uses AI, which would allow the relevant authorities to segment the area and gives insightful information for precautionary measures as well.

1.11.1. Definitions of the Basic Terms

Crowdsourcing: 'The use of IT to outsource any organizational function to a strategically defined population of human and non-human actors in the form of an open call' (Kietzmann, 2017).

Table 10. Average of Cumulative Score of VPOs and VPI.

Union Council	Outdoor Advertisements	Solid Waste	Hanging Wires	Dilapidated Buildings	Sewerage/ Drainage	Giraffiti	Poles/ Transformers	Encroachments	Broken Roads	Architecturally Poor Structures	Average of VPI
Afand colony	0.1910	0.0183	0.0631	0.0061	0.0288	0.0013	0.0443	0.0158	0.0053	0.0008	0.3746
Amar pura	0.1447	0.0434	0.0318	0.0073	0.0186	0.0015	0.0430	0.0140	0.0033	0.0009	0.3085
Asghar mill scheme	0.1883	0.0275	0.0497	0.0091	0.0103	0.0053	0.0446	0.0151	0.0041	0.0012	0.3551
Bangish colony	0.2010	0.0549	0.0536	0.0182	0.0206	0.0038	0.0450	0.0161	0.0081	0.0024	0.4236
Cantonment	0.1873	0.0529	0.0587	0.0292	0.0278	0.0034	0.0450	0.0166	0.0108	0.0043	0.4359
Chah sultan	0.1179	0.0220	0.0414	0.0073	0.0082	0.0015	0.0327	0.0193	0.0033	0.0009	0.2545
Chahmanzar colony	0.2010	0.0549	0.0536	0.0182	0.0206	0.0038	0.0450	0.0161	0.0081	0.0024	0.4236
Chaklala	0.1698	0.0682	0.0601	0.0315	0.0144	0.0026	0.0400	0.0151	0.0139	0.0017	0.4172
City	0.1924	0.0314	0.0655	0.0490	0.0184	0.0068	0.0450	0.0193	0.0080	0.0078	0.4437
Dha	0.2010	0.0549	0.0536	0.0182	0.0206	0.0038	0.0450	0.0161	0.0081	0.0024	0.3454
Dhoke ali akbar	0.0921	0.0000	0.0814	0.0000	0.0234	0.0147	0.0450	0.0209	0.0128	0.0000	0.4236
Dhoke babu irfan	0.1833	0.0469	0.0570	0.0191	0.0173	0.0024	0.0449	0.0182	0.0075	0.0015	0.2903
Dhoke dalal	0.1658	0.1167	0.0601	0.0270	0.0103	0.0019	0.0450	0.0150	0.0128	0.0012	0.3981
Dhoke farman ali	0.2010	0.2380	0.0703	0.0000	0.0520	0.0000	0.0450	0.0127	0.0000	0.0000	0.4559
Dhoke hukam dadad	0.0804	0.1309	0.0629	0.0000	0.0000	0.0000	0.0450	0.0127	0.0000	0.0000	0.6190
Dhoke kashirian	0.1039	0.0000	0.0888	0.0000	0.0000	0.0000	0.0296	0.0152	0.0000	0.0000	0.3319
Dhoke khabba	0.0972	0.0000	0.0537	0.0386	0.0000	0.0000	0.0315	0.0127	0.0000	0.0000	0.2374
Dhoke munshi	0.1734	0.0547	0.0435	0.0118	0.0145	0.0034	0.0418	0.0151	0.0058	0.0019	0.2336
Dhoke najjo	0.1764	0.0183	0.0610	0.0061	0.0253	0.0013	0.0450	0.0201	0.0027	0.0008	0.3660
Fauji colony	0.1916	0.0701	0.0518	0.0146	0.0164	0.0030	0.0450	0.0162	0.0065	0.0019	0.3571
Gangal	0.1491	0.0515	0.0484	0.0132	0.0133	0.0040	0.0401	0.0152	0.0051	0.0008	0.4171
Ganjmandi	0.1613	0.0157	0.0550	0.0257	0.0148	0.0011	0.0342	0.0221	0.0023	0.0046	0.3406

Girja	0.1005	0.1428	0.0648	0.0856	0.0338	0.0198	0.0450	0.0127	0.0175	0.0122	0.3368
Grand total	0.1718	0.0534	0.0533	0.0182	0.0198	0.0038	0.0423	0.0161	0.0078	0.0024	0.3889
Hazara colony	0.1910	0.0000	0.0518	0.0000	0.0000	0.0000	0.0231	0.0127	0.0000	0.0000	0.5346
Kartar pura	0.1826	0.0458	0.0447	0.0152	0.0171	0.0031	0.0438	0.0155	0.0068	0.0020	0.2786
Khayaban-e-sirsyed (n)	0.0704	0.0000	0.0333	0.0000	0.0468	0.0000	0.0450	0.0139	0.0233	0.0000	0.3765
Khayaban-e-sirsyed (s)	0.2010	0.1547	0.0666	0.0000	0.0000	0.0000	0.0450	0.0127	0.0000	0.0000	0.2327
Khuram colony	0.0704	0.0000	0.0518	0.0000	0.0520	0.0345	0.0450	0.0101	0.0000	0.0000	0.4800
Kotha kalan	0.1959	0.0585	0.0560	0.0178	0.0195	0.0054	0.0450	0.0162	0.0108	0.0028	0.2638
Lakhan	0.2010	0.0549	0.0536	0.0182	0.0206	0.0038	0.0450	0.0161	0.0081	0.0024	0.4280
Milat colony	0.1619	0.0183	0.0376	0.0061	0.0069	0.0013	0.0359	0.0151	0.0027	0.0044	0.4236
Mohallah imam bara	0.0905	0.0952	0.0518	0.0607	0.1040	0.0000	0.0390	0.0127	0.0280	0.0126	0.2901
Mohanpura	0.1943	0.0275	0.0601	0.0422	0.0380	0.0019	0.0450	0.0213	0.0146	0.0079	0.4944
Muslim town (e)	0.1686	0.1904	0.0715	0.0000	0.0347	0.0000	0.0344	0.0152	0.0000	0.0045	0.4528
New katarian	0.2010	0.0525	0.0530	0.0315	0.0224	0.0025	0.0443	0.0192	0.0054	0.0016	0.5193
Pindora	0.2010	0.0549	0.0536	0.0182	0.0206	0.0038	0.0450	0.0161	0.0081	0.0024	0.4333
Pir wadhai	0.1094	0.0793	0.0681	0.0000	0.0555	0.0115	0.0450	0.0207	0.0288	0.0000	0.4236
Purana qila	0.1340	0.0500	0.0610	0.0245	0.0069	0.0013	0.0354	0.0189	0.0027	0.0047	0.4183
Qayyum abad	0.1404	0.0738	0.0463	0.0000	0.0192	0.0066	0.0381	0.0136	0.0093	0.0000	0.3394
Ratta amral	0.1709	0.1131	0.0703	0.0442	0.0520	0.0000	0.0334	0.0146	0.0128	0.0000	0.3472
Sadiq abad	0.1441	0.0000	0.0925	0.0828	0.0468	0.0000	0.0450	0.0177	0.0257	0.0000	0.5112
Saidpur scheme	0.0771	0.0000	0.0296	0.0000	0.0000	0.0000	0.0450	0.0127	0.0000	0.0000	0.4546
Satellite town	0.1925	0.0449	0.0539	0.0149	0.0168	0.0059	0.0431	0.0155	0.0067	0.0019	0.1643
Shakaryal(janobi)	0.1128	0.0873	0.0456	0.0000	0.0416	0.0000	0.0362	0.0131	0.0117	0.0000	0.3961
Shakaryal(shamali)	0.0000	0.1071	0.0574	0.0000	0.0520	0.0000	0.0328	0.0165	0.0105	0.0000	0.3483
Waris khan	0.1675	0.0375	0.0402	0.0335	0.0051	0.0009	0.0394	0.0135	0.0020	0.0056	0.2762

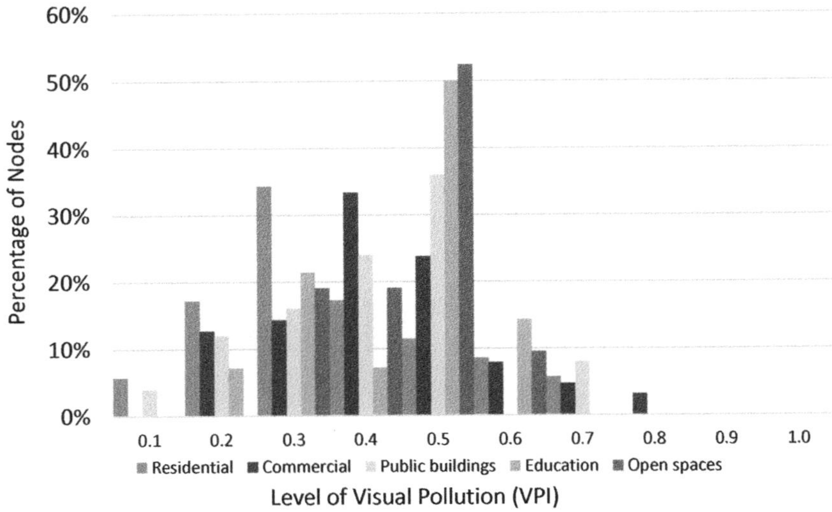

Fig. 30. Histogram of Visual Pollution Distribution vs. Land Uses.

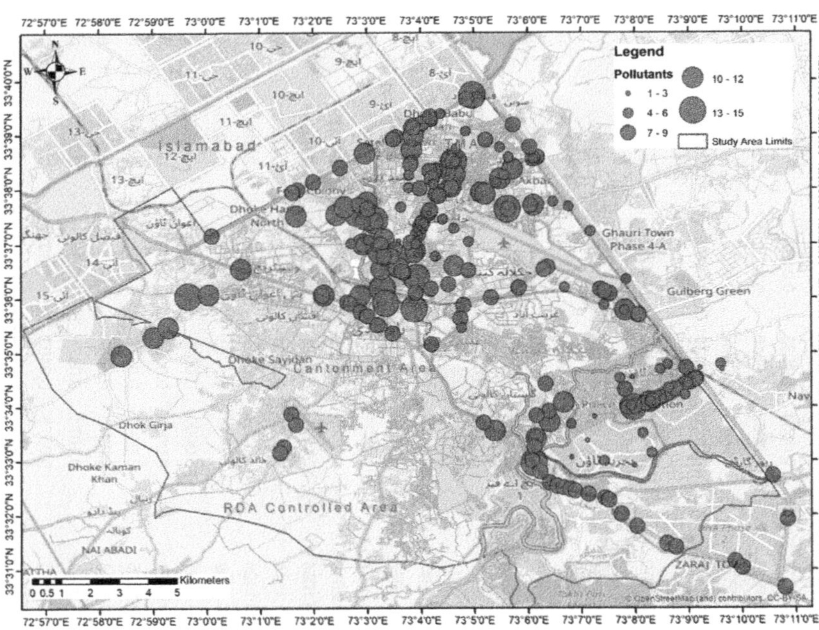

Fig. 31. Scaled Symbols Representation of Presence of VPOs.

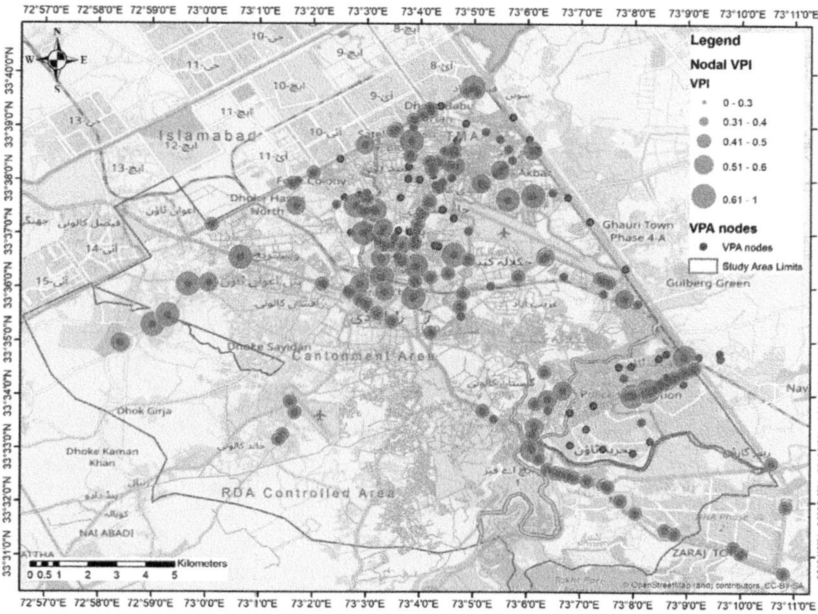

Fig. 32. Node Level Visual Pollution Score.

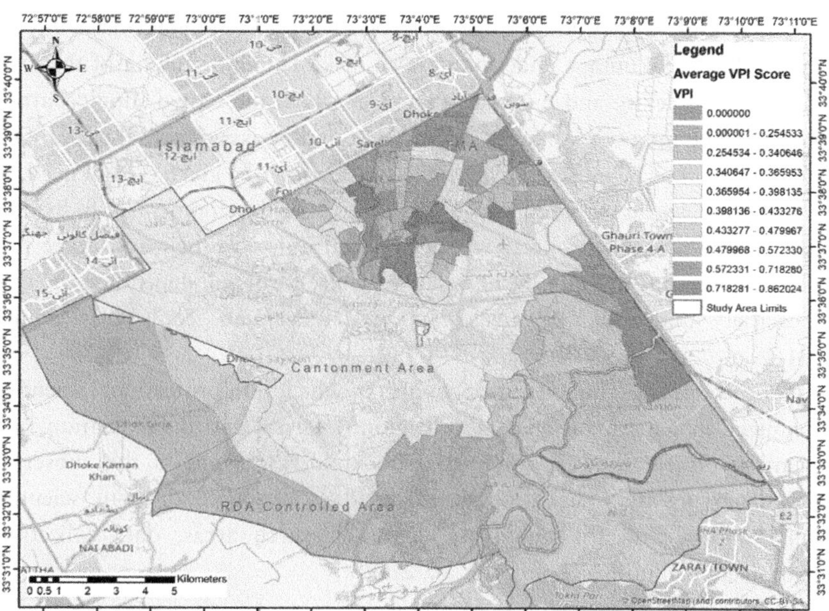

Fig. 33. Average VP Score Generalised at Union Council Level.

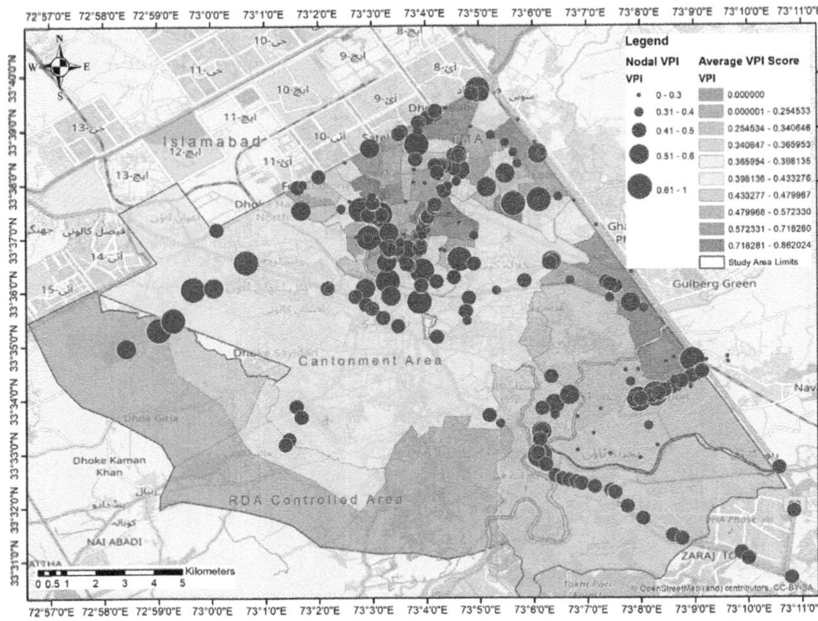

Fig. 34. UC Level VPI Viz-à-viz Nodal VPI.

'It is a type of participative online activity in which an individual, an institution, a nonprofit organization, or company proposes to a group of individuals of varying knowledge, heterogeneity, and number, via a flexible open call, the voluntary undertaking of a task. The undertaking of the task, variable complexity and modularity, and in which the crowd should participate bringing their work, money, knowledge, and/or experience, always entails mutual benefit. The user will receive the satisfaction of a given type of need, be it economic, social recognition, self-esteem, or the development of individual skills, while the crowdsourcer will obtain and utilize to their advantage that what the user has brought to the venture, whose form will depend on the type of activity undertaken' (Garrigos-Simon, Gil-Pechuán, & Estelles-Miguel, 2015).

Citizen science: The scientific activities in which non-professional scientists volunteer to participate in data collection, analysis and dissemination of a scientific project (Sui, Elwood, & Goodchild, 2013). It is the subset of crowdsourcing that has also been defined as a source of joint scientific collaboration with the public to identify research areas, data collection, analysis and solutions (Bonney et al., 2009).

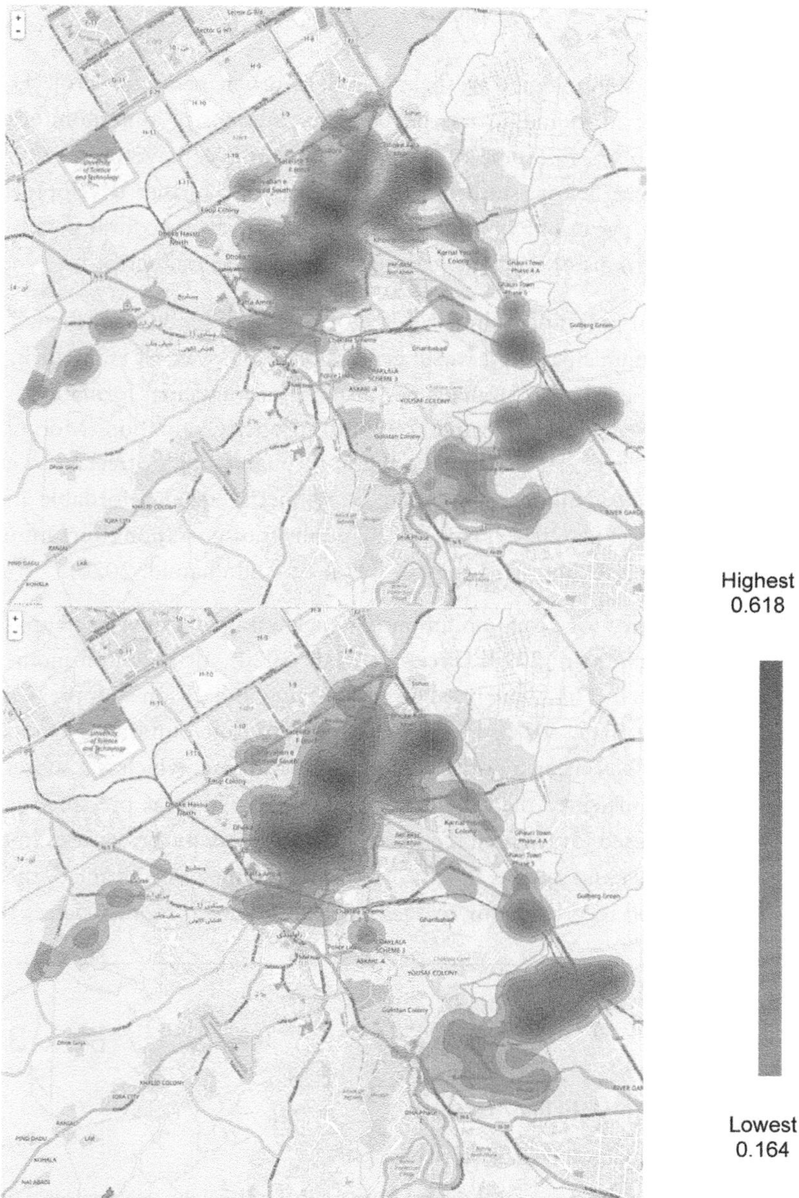

Fig. 35. Heat Map of Visual Pollution in the Study Area.

Artificial intelligence: It uses computers to perform those tasks that require human intelligence (Nadimpalli, 2017). Further, the study of mental faculties through computer language is called artificial intelligence (Charniak & McDormott, 2009).

1.11.2. Examples of Application of Modern Technologies

These emerging technologies are being used in urban areas for projects of various natures, including urban governance, administration, monitoring, evaluation and research. Some of the research examples have been referred from the literature below, showing their potential for further developments regarding visual pollution. In addition, the perspective procedural details of one of the algorithms of machine learning have also been described.

(1) AI for visual pollution (Ahmed, Islam, Tuba, Mahdy, & Sujauddin, 2019), for municipal solid waste generation (Abbasi & El Hanandeh, 2016; Wilts et al., 2021), machine learning for municipal formation (Kazakov, Kulagina, & Azarenko, 2020), property valuation (Morano, Tajani, & Torre, 2015), visual pollution through Google Street View and Yolo (Hossain, Nijhum, Sadi, Shad, & Rahman, 2021), affordable housing (Hu et al., 2019; Phan, 2019), public transportation (Bin Othman & Tan, 2019; Skhosana, Ezugwu, Rana, & Abdulhamid, 2020).

(2) Citizen science for visual pollution (Chmielewski, Samulowska, Lupa, Lee, & Zagajewski, 2018; Davies & Mah, 2020), air pollution monitoring (Castell et al., 2021) and crowdsourcing in decision support (Chiu, Liang, & Turban, 2014). Image Classification: Convolutional Neural Network (CNN) is an algorithm of machine learning, which is a subset of AI. CNN is used for the classification of images, whereas pre-labelled images are used for training the model. The CNN maps the features found in the images and assigns it the corresponding label; once trained, the CNN is used to predict any images/videos/live stream to auto-detect the

Box 3

Suggestions for labelled datasets

Some possible options for labelled datasets are:

- Local images of streets from different sites is used to label the environment with the given VPI, and the gathered dataset is used to train the CNN.
- Images using SAR or Google Images are used to train the network on a larger scale for larger VPO to further classify and categorise sites/cities accordingly.

corresponding objects. The same CNN could be used to predict the areas, sites or cities with high VPO, depending on what labels (Box 3) or dataset is used to train the network.

Once the CNN is trained, it could be deployed at different sites either within the city or at a larger scale. The deployed model will allow the model to classify objects with the given VPI. The outputs could provide sufficient information for management or regulatory bodies to take relevant actions, or the output itself could be utilised for another simple Machine Learning Linear Regression (LR). LR is simply the relationship between two or more variables.

The output information from the previous model will allow the authorities to gather sufficient information about the areas with high VPI and based on other generic information e.g. population, employment, education percentage or other attributes, which could help the LR model to create a more meaningful relationship between these variables. The relationship between these variables would allow municipalities to take more insightful action concerning the installation or maintenance of their sites.

The given output could also be used with other relevant information e.g. traffic volumes, busy times and location to create another simple LR model. The model would allow marketing teams to invest more sensibly and deploy their billboards and other outlets in a more effective way rather than crowd every site with multiple billboards without any profitable change.

There is multiple other applications of AI, which could be discussed depending on the problem and site. For example, in some areas, traffic volume is considered a VPO; hence machine learning could identify rush hours to avoid such sites. Also, the public transport could utilise this information to effectively layout their transport without affecting the audience, hence reducing the VPO within the environment.

1.11.3. Benefits and Risks of Modern Technologies

Table 11 presents the potentials, challenges and expected contribution of emerging technologies (Section 4.10) for the assessment of visual pollution for future interventions:

Table 11. Opportunities and Challenges of Modern Technologies for Visual Pollution Assessment.

Sr. No.	Potential Role in the VPA	Opportunities	Challenges
Artificial Intelligence	• It can be used to pick small details of VPOs with precision. • It can be applied for the bulk collection of VPOs data. • It can be worked	• Accuracy and precision • Deals with repetition • Time-saving • Monitoring • Enhancement of standardisation • Replacement of manned activities at dangerous locations (Nadimpalli, 2017)	• Data hungry system may lead towards inaccurate decisions • Unemployment of humans as the machines take over the most activities. It may cause difficulty for human capital. • The threat to human intelligence is stealing away their confidence in managing situations (Nadimpalli, 2017). • Fear of losing emotional strength and attachment as the dependency on the machines and computer system may increase.
Crowdsourcing	• It can be helpful for the data collection on VPOs' presence and characteristics using VPA scorecard. • This source can be beneficial for monitoring VPO status regarding quality and quantity. • In developing countries, it can be a way to involve youth for such development	• Building on the scientific knowledge by being part of the scientific community • Accessible way for data collection and public participation of relevant stakeholders • Development, design and encouragement of open data tools and sources	• Hidden and additional investments required for the success of projects • Expert team and management for engaging public, collecting and monitoring data • It may discourage skilled and quality suppliers because crowdsourcing brings the design

Table 11. *(Continued)*

Sr. No.	Potential Role in the VPA	Opportunities	Challenges
Citizen Science	works by the planning agencies and development authorities.	• Collaborative exchange of information, dissemination, suggestions and feedback • Users owned and users lead initiatives (Rouse, 2010) • Bring collective intelligence to the ideas, initiatives and projects • Engagement of scientific community • Quick and wide data collection approach	and other standards low.

1.12 Conclusion of the Chapter

The research and concern on the collective assessment of visual pollution have been started and the first systematic, objective approach and android-based VPA tool have been introduced. But still, the further involvement of modern technologies, particularly to gauge the level of visual pollution, needs scientific and applied research. Especially for poor and weak economies, what can be the best strategies to adopt such technology-driven initiatives? The benchmark and limits for VPI varies for developing and developed countries as the level of urban governance, socioeconomic, infrastructure and services are not the same between them. That again seeks more comprehensive, comparative and geographic-specific studies to uniform the hypothesis.

1.13 Summary of the Chapter

(1) The quantification of visual pollution has always been challenging due to its high subjectivity and the absence of one structured mechanism/tool/approach like air, water, noise and land pollution.

(2) In the recent past, a new Open Data Kit-based VPA tool has been developed and published by the authors to calculate visual pollution levels at any desired location.

(3) It consists of two parts: VPA scorecard and VPA calculator. VPA scorecard collects data on VPOs presence and their characteristics and, after processing, assigns a numeric score.

(4) The real-time application of the tool has been performed at 303 sampled nodes in one of the urban areas of Pakistan having different strata of the socioeconomic population, with diverse types of land uses.

(5) The results of VPA tool can be visualised as heat maps by scaling the visual pollution score on a scale of 0–1. The author used Heatmapper, an automated tool.

(6) The book adds new definitions and terminologies of visual pollution, VPOs, VPA and VPI to the current body of literature related to the subject.

(7) The quantification and assessment of visual pollution can be enhanced by experimenting with modern technologies including AI, crowdsourcing and citizen science conditioned to the availability of prerequisites discussed in Section 4.10.

5

MANAGEMENT AND MITIGATION OF VISUAL POLLUTION

1. INTRODUCTION

The book's last section concludes the previous chapters and proposes the five-tier legal and administrative framework for regulating visual pollution. It suggests the following five dimensions to address the formulation of practical actions and reforms. All these sectors require parallel and sequential tasks on the following agenda.

(1) Knowledge building on visual pollution.

(2) Ensuring awareness amongst planning professionals at the academic and administrative levels.

(3) Formulation of a legal framework to address visual pollution.

(4) Quantification and mapping of visual pollution in urban centres.

(5) Remedial measures for visual pollution (by identifying temporary, short-term and long term visual pollutants).

The proposed framework includes many stakeholders by citing international examples to achieve the desired level of implementation regarding visual pollution. It includes administrative and public bodies, non-government organisations (NGOs), private developers, academia and communities to ensure a synchronised and monitored system against visual pollution. Further, this chapter suggests future research areas for improved application of visual pollution objects (VPOs) and the visual pollution assessment (VPA) tool for assessment and spatial quantification of visual pollution alongside identification of areas on visual pollution which must be included in environmental policies and legal framework to address the issue pragmatically.

1.1 Policy and Administrative Management Frameworks

The discussion on case studies and assessment methods related to visual pollution in Chapters 3 and 4 shows the difference in management practices between developing and developed countries. It is further concluded in the previous chapters that the problem of visual pollution is linked with urbanisation that needs to be dealt with primarily through policy initiatives. Chapter 3 also presents the weaknesses of policy and management practices of the institutions governing visual pollution of different countries. A cohesive multidimensional approach is required to effectively manage and control visual pollution. Therefore, the following five-pillar framework can be adopted to design practical actions and reforms (Fig. 36).

1.1.1. Knowledge Building on Visual Pollution

The current status of the published literature in English and its skewness in different categories, as described in Chapters 2 and 3, indicate that this branch of knowledge is evolving. Further, it presents that the visual pollution problem has not been fully realised by developing countries' planning institutions, urban planners and related professionals. There is a dire need to bridge knowledge gaps in the three categories: (1) revising contents of urban design related academic curriculum, (2) developing user-friendly modules for the

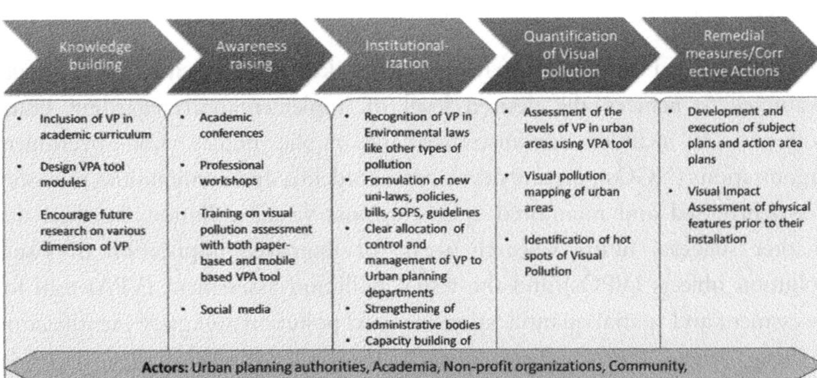

Fig. 36. Visual Pollution Control and Management Framework.

widespread application of VPA tool and (3) motivating researchers to pursue future research in the field of visual pollution.

First, there is a severe need to revise the content of the academic curriculum concerning environmental laws, planning, and practices. Presently, there is no single course covering the topic of visual pollution, which would be the initial step towards a structured educational plan about the issue. Second, the necessity of exploring visual pollution related research areas will further evolve with the development of fundamental comprehension of the subject by academia. In the near future, the researchers may focus on exploring:

- How to replicate and improvise the VPA tool in other study areas across the globe?
- What will be the content of new environmental laws, policies etc. for visual pollution and how will that be incorporated into existing legislation?
- What challenges would quantification and mapping of visual pollution face in urban areas, particularly megacities?
- How can machine learning and artificial intelligence be used to assess visual pollution both in planned and unplanned areas?

1.1.2. Awareness Promotion Among Planning Professionals

The research indicates that visual pollution is a less-recognised area of study, particularly by administrative bodies, practitioners and researchers. This section identifies some key activities to increase the comprehension of the concept, details of visual pollution and addressing the inherent barriers.

It is thus highly recommended: to conduct intensive training of the concerned professionals, especially urban planners and managers, to connect them with the new mandate of visual pollution; capacitate them to quantify the threshold of visual pollution through VPA tool and remove the reluctance in using the mobile-based VPA tool. Further, it is highly proposed to have extensive and periodic training sessions for the teaching faculties, practitioners, managers, policymakers and urban planning students on visual pollution. Here are some crucial channels to escalate the understanding of stakeholders about the subject of visual pollution:

- *Academic conferences (physical and online):* These can be organised by concerned colleges, universities and departments of urban planning, urban design, architecture or built environment etc.

- *Professional workshops (physical and online):* It is recommended that all stakeholders, including planning departments, environmental agencies, architectural and planning firms, civil society etc. should arrange hands-on seminars, focus group discussions and symposiums on the utilisation, application of VPA tool and visualisation of spatial heat maps with red zones of visual pollution.
- *Social media:* Multiple social media based mediums can be utilised to generate discussions among the professional communities about the issue. These are accessible and convenient means of communication, which, if used tactfully, can facilitate rapid developments in visual pollution.
- *Print media (newspapers, magazines):* Historically, this was one of the first forms of communication used by researchers and columnists to raise voice against visual pollution. Today, it is considered the same. Articles and columns by the urban planners, urban designers, environmentalists, city managers and researchers can generate future discussions and build pressure groups for safe, clean and litter-free urban environments.
- *Training on the use of VPA mobile-based:* the research presents the VPA tool for recording the number of VPOs and their characteristics at any location, ultimately quantifying the visual pollution at a scale of 0–1 to view the heat maps of urban settlements. The VPA tool is a new, robust and tested methodology to estimate the level of visual pollution at the desired point. The paper and mobile-based versions of the tool arise from conducting training sessions and forums to appropriately guide and capacitate end users.

1.1.3. The Institutionalisation of Visual Pollution in Legal Framework

Revision of existing legal documents and strengthening physical and human resources must be prioritised to promote awareness about visual pollution. The environmental laws, policies, inking by-laws, development control practices and standard operating procedures (SOPs) about all VPOs require meaningful additions, alterations and reforms to introduce uniform development. Another problem is the inability to implement policies and effective management by the concerned authorities. It is imperative to strengthen the current institutions with legal framework through:

- Recognition of visual pollution in environmental acts and laws like any other type of pollution

- Formulation of new laws, policies, bills, SOPs, guidelines, beautification acts covering all visual pollution aspects
- Strengthening of administrative bodies by capacity building of relevant professionals and providing required resources to urban planning departments in collaboration with communication networks, telecom agencies, construction and engineering units etc.

1.1.4. Remedial Measures/Corrective Actions for Visual Pollution

This part of the management and control framework for visual pollution requires the implementation of the following steps.

- Development and execution of subject plans and action area plans
- Visual impact assessment (VIA) of physical structures before their installation

This would require the identification of most hazardous areas of visual pollution after quantification of its levels using VPA tool. The categorisation of locations will help administrative bodies take immediate corrective actions in compliance with the visual aesthetics goals of the urban environments. It will also assist the identification of the type of VPO which needs to be treated first. The emergency plans may be of varying nature like: removal of unethical content of billboards/hoardings and its replacement with some public messages, removal of illegal billboards, clearing of encroachments from the markets and residential areas, regularisation of sign boards of shops and offices and beautification of existing structures.

VIA of physical features before their installation stands for carrying out a VIA of various physical elements like billboards, hoardings, bins of solid waste, transformers of electricity, poles of communication etc. before their installation to check the feasibility and suitability of that particular feature at some specific site. The VIA will have the same mandate and value as of Traffic Impact Assessment (TIA) of Environmental Impact Assessment (EIA), which assesses the impact of any new intervention on the existing traffic flows and the quality of the environment in the future, respectively. Such visual impact assessment is in practice in developed countries.

1.1.5. Quantification and Mapping of Visual Pollution in Urban Centres

The fifth step of the visual pollution management framework is preparing baseline datasets for the density of visual pollution levels using the VPA tool and its spatial visualisation in growing urban areas. The legislative provisions and reforms regarding visual pollution as a new type of pollution will force the planning agencies to execute requisite tasks. The sequential steps to be carried out for the assessment and mapping of visual pollution include:

- Listing of urban areas
- Preparation of base data layers for all urban areas (single-line road network, the polygon of urban blocks, land use, city limits and boundaries)
- Identification of sampled points
- Data collection on sampled points using VPA tool
- Calculation of visual pollution score and Index
- Determination of visual pollution threshold category on sampled points
- Spatial mapping of visual pollution threshold on sampled points
- Declaration of heat maps of visual pollution

1.2 Laws of Visual Pollution

Based on the discussion and analysis of the previous chapter, the following laws and basics (Box 4) have been drawn for visual pollution:

Box 4

ABC of Visual Pollution

- Population awareness
- Beautifying surroundings
- Progressively conditioning stakeholders to better environment
- Connecting with nature, coordinating with institutions, capacitating professionals
- Decluttering surroundings
- Respecting cultures
- Maintaining order
- Quantifying visual pollution

- *First Law:* Visual pollution is inversely proportional to the income and education level of the residents and strictness of the governing authorities
- *Second Law:* The spread of visual pollution has a higher pace in commercial land use.
- *Third Law:* Planned human settlements have lesser visual pollution than corresponding unplanned settlements

1.3 Conclusion of the Chapter

The management of visual pollution is a multispectral and parallel phenomenon that can be executed efficiently through an effective management framework. In a nutshell, it can be concluded that controlled management of visual pollution = good urban governance + strict urban development + political will + requisite legislation.

1.4 Summary of the Chapter

(1) The management of visual pollution is a five-pillar subject that needs parallel and sequential inputs. These pillars include knowledge building of the subject, awareness building of professionals, formulation and revision of related laws, execution of remedial actions and quantification of visual pollution in urban settlements coupled with spatial mapping.

(2) The knowledge building of the subject demands further research both on theoretical and practical grounds. The content of visual pollution from developing countries can showcase the complex flavour of this problem and can be further employed to compare the cases of visual pollution between developed and developing countries.

(3) Public and professional level comprehension and recognition are basic tasks that can be attained through awareness campaigns on social media, print media, local symposiums, international conferences, dialogues, discussion and academic training.

(4) Different case studies (Chapter 3) are proof that surgical actions against visual pollution with legal support have produced results. Devising new policies and revising existing rules and regulations are an absolute must for institutional development to combat visual pollution.

(5) Remedial actions consist of identifying subject plans for visually critical areas and implementing corrective strategies for them. It also includes the concept of VIA that has to be executed before installing any physical elements in the urban settings.

(6) The assessment of the visual pollution with heat maps spatial visualisation using the VPA tool can facilitate better situational analysis.

(7) The involvement and engagement of smart technologies and advancements like artificial intelligence, crowd sourcing and citizen science to explore this subject can be a potential future research area.

BIBLIOGRAPHY

Abbasi, M., & El Hanandeh, A. (2016). Forecasting municipal solid waste generation using artificial intelligence modelling approaches. *Waste Management*, *56*, 13–22. doi:10.1016/j.wasman.2016.05.018

Abdelhamid, M. M. (2018). An attempt to reduce visual pollution in the building sector within Egyptian cities. *Spaces and Flows: An International Journal of Urban and ExtraUrban Studies*. Retrieved from http://search.ebscohost.com/login.aspx?direct=true&profile=ehost&scope=site&authtype=crawler&jrnl=21548684&AN=136111844&h=M7nkngiEkrFWaO%2FxFo4Tq8SZZdZA1ZPxrhWTkF%2BSsKqwHeoj3OTeqRZgI6Z5zjivyVFPBQzHadV9s%2Bt0460bpA%3D%3D&crl=c

Ahmed, N., Islam, M. N., Tuba, A. S., Mahdy, M. R. C., & Sujauddin, M. (2019). Solving visual pollution with deep learning: A new nexus in environmental management. *Journal of Environmental Management*, *248*, 109253. Retrieved from https://www.sciencedirect.com/science/article/pii/S0301479719309557

Al-Anbari, M. A., Abd, M. A., Obaid, A. H. (2020). Evaluation of some visual pollution indicators in the physical environment of Al-Hilla city. *Muthanna Journal of Engineering and Technology*. Retrieved from http://engineering.uobabylon.edu.iq/fileshare/articles/repository1_publication6158_15_227.pdf

Al-Sharaa, A., Bakar, S. A., & Maulan, S. B. (2016). Determining peoples threshold towardes visual pollution using areal cumulative analysis and public survey. In MLA 2016 committee. Retrieved from http://mlaupm.com/an-component/media/download/MLA_SEMINAR_2016.pdf#page=18

Allahyari, H., Nasehi, S., Salehi, E., & Zebardast, L. (2017). Evaluation of visual pollution in urban squares, using SWOT, AHP, and QSPM techniques (Case study: Tehran squares of Enghelab and Vanak). *Pollution*, *3*(4), 655–667. doi:10.22059/POLL.2017.62780

Allen, M. (2017). *The SAGE encyclopedia of communication research methods*. doi:10.4135/9781483381411NV-4

Amin, K. (2002). Urban quality and designing of spaces: Case study for Nasr city, Cairo. In 38th international planning congress (pp. 1–11).

Anciaes, P. R. (2014). Book review: Visual pollution: Advertising, signage and environmental quality by Adriana Portella. *LSE Review of Books*. Retrieved from http://eprints.lse.ac.uk/57500/

Anderson, P. M. L., Okereke, C., Rudd, A., & Parnell, S. (2013). Regional assessment of Africa, *Urbanization, biodiversity and ecosystem services: Challenges and opportunities: A global assessment*. Springer. doi:10.1007/978-94-007-7088-1_23

Anwer, M., & Aowais, A. (2019). Visual pollution and its impact on the aesthetics: The town of Eizariya as a model. *Journal of the Planner and Development*. Retrieved from https://www.iasj.net/iasj/article/184101

Atta, H. A. (2013). Visual pollution and statistical determination in some of Karrada district main streets, Baghdad. *Journal of Engineering*, *19*(3), 414–428.

Azeema, N. (2015). Billboard advertisement visual pollution. (Dr. Bernard Weitzman).

Azeema, N., & Nazuk, A. (2016). Is billboard a visual pollution in Pakistan? *International Journal of Scientific Engineering and Research*, *7*(7), 862–874.

Babcock, H. (2015). Is using the public trust doctrine to protect public parkland from visual pollution justifiable doctrinal creep. *Ecology LQ*, *42*, 1.

Baesse, S. (2012). *Towards more effective urban planning in Jeddah, Saudi Arabia (issue February)*. RMIT University, Melbourne School of Global Studies, Social Science and Planning. Retrieved from https://researchbank.rmit.edu.au/eserv/rmit:161451/Baesse.pdf

Bakar, S. A., al-Sharaa, A., Maulan, S., & Munther, R. (2019). Measuring visual pollution threshold along Kuala Lumpur historic shopping district streets using cumulative area analysis. digitalcommons.esf.edu. Retrieved from https://digitalcommons.esf.edu/vrconference/16/

Bakhotmah, F. A. (1998). Visual amenity & visual pollution definition.

Banerjee, S. (2017). A study of visual pollution and its effect on mental health. *Scholarly Research Journal for Interdisciplinary Research*. Retrieved from https://www.academia.edu/download/53256948/13.Mrs._Sudeepta_Banerjee_1.pdf

Bankole, O. (2013). Urban environmental graphics: Impact, problems and visual pollution of signs and billboards in Nigerian cities. *International Journal of Education and Research*, *1*(6), 1–12.

Barrett, J. H. (1972). Visual pollution in a national park. *Marine Pollution Bulletin*. Retrieved from https://www.sciencedirect.com/science/article/pii/0025326X72902147

Beardsley M. C. (1982). Aesthetic welfare, aesthetic justice, and education Policy. In M. J. Wreen & D. M. Callen (Eds.), *The aesthetic point of view: Selected essays*. M. C. Beardsley (pp.111–124). London: Cornell University Press.

Bedin, B., Ferrari, M., & Gajardo, R. (2015). Visual pollution and its control in Caxias do Sul county from the municipal law no. 412/2012/A Poluicao visual Eo Seu Controle no .412/2012. *Direito Da Cidade*. Retrieved from https://go.gale.com/ps/i.do?id=GALE%7CA566559048&sid=googleScholar&v=2.1&it=r&linkaccess=abs&issn=23177721&p=IFME&sw=w

Beer, C. (2016). Adriana Portella: Visual pollution: Advertising, signage and environmental quality. *Human Ecology*. Retrieved from http://search.proquest.com/openview/3951d14c52b0aa6de72c21440f46f311/1?pq-origsite=gscholar&cbl=48275

Benavides, D. A. M. (2017). Environmental diagnosis of visual pollution perception by university population from the civil engineering and natural, exact and education sciences faculties at Universidad del Cauca. *Luna Azul*. Retrieved from http://www.scielo.org.co/scielo.php?pid=S1909-24742017000100013&script=sci_arttext&tlng=en

Bi-xuan, M. (2014). On exploration for management and strategies of visual pollution in historic streets. *Shanxi Architecture*. Retrieved from http://en.cnki.com.cn/Article_en/CJFDTotal-JZSX201416011.htm

Bin Othman, M. S., & Tan, G. (2019). Machine learning aided simulation of public transport utilization. In Proceedings of the 2018 IEEE/ACM 22nd international symposium on distributed simulation and real time applications, DS-RT 2018 (pp. 253–254). doi:10.1109/DISTRA.2018.8601011

Blewitt, J. (2015). *Book review: Visual pollution: Advertising, signage and environmental quality*. Taylor & Francis. Retrieved from https://www.tandfonline.com/doi/full/10.1080/02665433.2014.967496

Bonney, R., Cooper, C. B., Dickinson, J., Kelling, S., Phillips, T., Rosenberg, K. V., & Shirk, J. (2009). Citizen science: A developing tool for expanding science knowledge and scientific literacy. *BioScience*, *59*(11), 977–984. doi:10.1525/bio.2009.59.11.9

Campos, C., & Cirafici, A. (2020). Visual pollution and social asymmetry. The origin of Dientenegro. *DISEGNARECON*. Retrieved from http://disegnarecon.univaq.it/ojs/index.php/disegnarecon/article/view/731

Caprotti, F., Cowley, R., Datta, A., Broto, V. C., Gao, E., Georgeson, L., … Joss, S. (2017). The new urban agenda: Key opportunities and challenges for policy and practice. *Urban Research and Practice*, *10*(3), 367–378. doi:10.1080/17535069.2016.1275618

Carrizo, J. G. (2016). Conference program NDiMR 2016. In City and communication: The need for sustainable outdoor advertising (p. 18).

Castell, N., Grossberndt, S., Gray, L., Fredriksen, M., Skaar, J. S., & Høiskar, B. A. K. (2021). Implementing citizen science in primary schools: Engaging young children in monitoring air pollution. *Frontiers in Climate*, *3*, 639128.

Cauduro, J. C. (1981). *Design & Ambiente (design and environment)*. Sao Paulo: FAUUSP.

Center for Local Governance. (2019). *Tackling urban visual pollution to enhance the Saudi cityscape* CLG: Riyadh, Saudi Arabia.

Cercleux, A.-L., Merciu, F.-C., & Merciu, G.-L. (2016). A model of development strategy encompassing creative industries to reduce visual pollution – Case study: Strada Franceză, bucharest's old city. *Procedia Environmental Sciences*, *32*, 404–411. doi:10.1016/j.proenv.2016.03.046

Charniak, E., & McDormott, D. (2009). *Introduction to Artificial intelligence – Eugene charniak – Google books (Fourth)*. Pearson Education. Retrieved from https://books.google.com.pk/books?hl=en&lr=&id=HACaS635bYcC&oi=fnd&pg=PA1&dq=what+is+artificial+intelligence&ots=Km3urUZK5S&sig=xc1jezFZYaYOQfFAcfEmNUXnyBg&redir_esc=y#v=onepage&q=what is artificial intelligence&f=false

Cheng-lin, H. (2003). On the problems of touring ecology destruction and visual pollution in the "two-mountain and one-lake area" of Anhui province [J]. *Journal of Catastrophology*. Retrieved from http://en.cnki.com.cn/Article_en/CJFDTotal-ZHXU200304015.htm

Chiu, C. M., Liang, T. P., & Turban, E. (2014). What can crowdsourcing do for decision support? *Decision Support Systems*, *65*(C), 40–49. doi:10.1016/j.dss.2014.05.010

Chmielewski, S. (2020). Chaos in motion: Measuring visual pollution with tangential view landscape metrics. *Land*, *9*(12), 515. doi:10.3390/land9120515

Chmielewski, S. (2021). Towards managing visual pollution: A 3D isovist and voxel approach to advertisement billboard visual impact assessment. *ISPRS International Journal of Geo-Information*, *10*(10). doi:10.3390/ijgi10100656

Chmielewski, S., Chmielewski, T. J., Tompalski, P., & Wezyk, P. (2011). Geo-spatial problem of advertising billboards in urban landscape: Lublin (Poland) case study. 1–3.

Chmielewski, S., Lee, D. J., Tompalski, P., Chmielewski, T. J., & Wężyk, P. (2016). Measuring visual pollution by outdoor advertisements in an urban street using intervisibilty analysis and public surveys. *International Journal of Geographical Information Science*, *30*(4), 801–818. doi:10.1080/13658816.2015.1104316

Chmielewski, S., Samulowska, M., Lupa, M., Lee, D., & Zagajewski, B. (2018). Citizen science and WebGIS for outdoor advertisement visual pollution assessment. *Computers, Environment and Urban Systems*, *67*(January), 97–109. doi:10.1016/j.compenvurbsys.2017.09.001

Choudhary, A. (2016). Model to mitigate visual pollution by ADS and signage. *4*(3), 516–521.

Choudhary, A., & Shrivastava, A. T. (2016). Model to mitigate visual pollution by ADS and signage. *International Journal of Engineering Research and General Science*, *4*(3), 516–521.

Chuan, W. (2009). Stop urban visual pollution. *Architecture & Culture*. Retrieved from http://en.cnki.com.cn/Article_en/CJFDTotal-JZYW200911020.htm

CityPopulation. (2017). Pakistan: Provinces, districts, subdistricts, cities, towns – Population statistics in maps and charts. Retrieved from http://www.citypopulation.de/Pakistan.html

Club, S. (1966). Visual pollution. Sierra club policies. Sierra club conservation policies.

Codato, M. V. F. (2014). Visual pollution and noise: A troubled relationship between environment and society. *Revista Eletrônica em Gestão Educação e Tecnologia Ambiental*. Retrieved from https://periodicos.ufsm.br/index.php/reget/article/view/14516

Condly, S. J. (2003). The psychologies of gestalt principles of visual perception and domain expertise: Interactions and implications for instructional practice. *Florida Association of Teacher Educators Journal*, *1*(3), 1–17.

Córdoba, L. Y. (n.d.). El entorno histórico de la ciudad de Cartagena opacado por la contaminación visual comercial. *Aglala*, *4*(1), 72–92.

Correa, V. F., & Mejía, A. A. (2015). Visual pollution indicators and its effects on population. *Enfoque UTE*. Retrieved from http://scielo.senescyt.gob.ec/scielo.php?script=sci_arttext&pid=S1390-65422015000300115&lng=en&nrm=iso&tlng=en

Council, J. C. (1985). Visual pollution study. Retrieved from https://digitalcommons.unf.edu/cgi/viewcontent.cgi?article=1003&context=jcci

Cullen, G. (1961). *The concise landscape*, edited by The Architectural Press (1st ed.). Routledge. Retrieved from https://www.routledge.com/Concise-Townscape/Cullen/p/book/9780750620185. Accessed on August 20, 2022.

Cunningham, R. A. (2016). Billboard control under the highway Beau Tification act of 1965. *Michigan Law Review*, *71*(7). Retrieved from http://www.jstor.org/stable/1287593

Cvetković, M., & Momčilović-Petronijević, A. (2018). Visual pollution of the historical city core–A case study, the city of NIŠ. In Proceedings of the 6th International conference contemporary achievements in civil engineering, Subotica, Serbia. Retrieved from http://zbornik.gf.uns.ac.rs/doc/NS2018.49.pdf

Cvetković, M., Momčilović-Petronijević, A., & Ćurčić, A. (n.d.). Visual pollution of urban areas as one of the main issues of the 21st century. Researchgate.Net. Retrieved from https://www.researchgate.net/profile/Mila_Cvetkovic3/publication/331718054_VISUAL_POLLUTION_OF_URBAN_AREAS_AS_ONE_OF_THE_MAIN_ISSUES_OF_THE_21ST_CENTURY/links/5c8bca3da6fdcc381755c205/VISUAL-POLLUTION-OF-URBAN-AREAS-AS-ONE-OF-THE-MAIN-ISSUES-OF-THE-21ST-C

Daluge, M. J., DeLong, M., Hanig, L., Kalla, H., Klauer, C., Klein, K., … Wessinger, B. (2011). *Outdoor advertising control practices in Australia, Europe, and Japan*. Retrieved from https://rosap.ntl.bts.gov/view/dot/1013

Davies, T., & Mah, A. (2020). *Toxic truths*. [T. Davies & A. Mah] (1st ed.). Manchester University Press. Retrieved from https://library.oapen.org/bitstream/handle/20.500.12657/42642/9781526137005_fullhl.pdf?sequence=1#page=173

Devisscher, T., Konijnendijk, C., Nesbitt, L., Lenhart, J., & Salbitano, F. (2020). *Sustainable cities and communities – impacts on forests and forest-based livelihoods*. doi:10.1017/9781108765015

Douglas, I. (1983). *The urban environment*. Hillsdale NJ: Hodder Arnold.

Dukic, T., Ahlstrom, C., Patten, C., Kettwich, C., & Kircher, K. (2013). Effects of electronic billboards on driver distraction. *Traffic Injury Prevention, 14*(5), 469–476. doi:10.1080/15389588.2012.731546

Dumitrescu, A. (2000a). Dams, visual pollution and environment damage. *Revista de Ecologie Industriala*, 7(10), 59–62.

Dumitrescu, A. (2000b). Wind turbines, solar panels and visual pollution. *Revista de Ecologie Industriala*, 7(10), 63–66.

Dunn, M. (2006, March). Educating for a sustainable community. Environmental topic: Visual pollution. Retrieved from http://www.cabq.gov/aes/s5vp.html

Edquist, J. (2009). *The Effects of visual clutter on driving performance* (p. 226). Monash University. Retrieved from https://www.tml.org/legal_pdf/Billboard-study-article.pdf

EL-Ghonaimy, I. H. (2019). Visual pollution phenomena and sensitivity of residences in heritage city centers case of: Old district of Manama city, Kingdom of Bahrain. *Journal of Contemporary Urban Affairs*. Retrieved from http://www.ijcua.com/index.php/ijcua/article/view/125

Elena, E., Cristian, M., & Suzana, P. (2012). Visual pollution: A new axiological dimension of marketing? *Annals of Faculty of Economics*, 1(2), 820–826.

Emeji, M. J. (2011). Environmental communication and visual pollution in the Nigerian city of port Harcourt: Implications for design education and city planning. *Visual Arts Research*. Retrieved from https://muse.jhu.edu/article/451821/summary

Enache, E., Morozan, C., & Purice, S. (2012). Visual pollution: A new axiological dimension of marketing? *Annals of the University of Oradea, Faculty of Economics*. Retrieved from http://search.ebscohost.com/login.aspx?direct=true&profile=ehost&scope=site&authtype=crawler&jrnl=15825450&asa=Y&AN=85949029&h=xXLZRQmlkXPDzjeuzvoSb9nThpntHN0hPTYznJkP5P83S6j76f0964al4%2Fjm%2FC5%2FuY8%2B3I83x5jP668pcGJzjA%3D%3D&crl=c

Erickson, L. E. (1977). Approach to valuing visual pollution from western electricity production. Retrieved from https://www.osti.gov/biblio/7307006

ESRI. (2020). Creating a point feature at an intersection—ArcMap | Documentation. *ArcGIS Desktop Documentation*. Retrieved from https://desktop.arcgis.com/en/arcmap/latest/extensions/production-mapping/creating-point-features-at-intersections.htm

Eyenga, I. I., Focke, W. W., Prinsloo, L. C., & Tolmay, A. T. (2002). Photodegradation: A solution for the shopping bag "visual pollution" problem? *Macromolecular Symposia*, *178*(1), 139–152.

Fischer, L., & Abllez, M. E. (1996). Campus communication or visual pollution? scholarworks.lib.csusb.edu. Retrieved from https://scholarworks.lib.csusb.edu/cgi/viewcontent.cgi?article=1396&context=coyote-chronicle

Flad, H. K. (1997). Country clutter: Visual pollution and the rural roadscape. *The Annals of the American Academy of Political and Social Science*, *553*(1), 117–129.

Furze, J. (2002). Stealth wind turbines: Designs and technologies to reduce visual pollution. *Refocus*. Retrieved from https://www.sciencedirect.com/science/article/pii/S1471084602800221

Galea, S., Ahern, J., Rudenstine, S., Wallace, Z., & Vlahov, D. (2005). Urban built environment and depression: A multilevel analysis. *Journal of Epidemiology & Community Health*, *59*, 822–827.

Garrigos-Simon, F. J., Gil-Pechuán, I., & Estelles-Miguel, S. (2015). Advances in crowdsourcing. *Advances in Crowdsourcing*, 1–183. doi:10.1007/978-3-319-18341-1

Gokhale, V. (2011). Examining impact of visual pollution on city environment: Case study of Pune, India. *Spandrel Journal of School of Planning and Architecture Bhopal*, *1*, 10.

Gokhale, V. A., & Mayuri Raichur, I. R. (n.d.). Examining impact of visual pollution on city environment: Case study of Pune, India.

Government of Pakistan, G. (2018). 6th population and housing census 2017 (provisional results). Pakistan Bureau of Statistics. Retrieved from http://www.pbscensus.gov.pk/

Grygny, N. (2014). Reklama w przestrzeni miejskiej visual pollution–zawłaszczanie i odzyskiwanie miasta. ruj.uj.edu.pl. Retrieved from https://ruj.uj.edu.pl/xmlui/handle/item/197984

Hakiminejad, A. (2018). Measuring cities: A study of the development of Iranian urban sustainability assessment mechanisms from a UK perspective [West London]. Retrieved from https://www.researchgate.net/figure/Arups-SPeAR-diagram-Arup-2012_fig6_325090090

Harzing, A. W. (2007). Publish or perish. Retrieved from http://www.harzing.com/pop.htm

Harzing, A. W. (2010). *The publish or perish book*. Melbourne: Tarma Software Research Pty Limited.

Hastaoglou-Martinidis, V. (2011). Urban aesthetics and national identity: The refashioning of eastern mediterranean cities between 1900 and 19401. *Planning Perspectives*, *26*(2), 153–182. doi:10.1080/02665433.2011.550442

Hemantha, W. (2014, December 25). CEJ case produced visual pollution regulation in Sri Lanka. Retrieved from https://hemanthawithanage.blogspot.com/2014/12/cej-case-produced-visual-pollution.html

Hogg, R. V. (2016). *Instructor's solutions manual probability and statistical inference*. Upper Saddle River, NJ: Prentice Hall.

Hogg, R. V., Tanis, E. A., & Zimmerman, D. L. (2010). *Probability and statistical inference*. Upper Saddle River, NJ: Pearson/Prentice Hall.

Holdstein, R. E. (1974). *The problem of visual pollution and its relation to the role of the art educator*. Columbus, OH: Ohio State University.

Hossain, M. Y., Nijhum, I. R., Sadi, A. A., Shad, M. T. M., & Rahman, R. M. (2021). Visual pollution detection using google street view and Yolo. In 2021 IEEE 12th annual ubiquitous computing, electronics & mobile communication conference (UEMCON) (pp. 0433–0440). doi:10.1109/UEMCON53757.2021.9666654

Hu, L., He, S., Han, Z., Xiao, H., Su, S., Weng, M., & Cai, Z. (2019). Monitoring housing rental prices based on social media: An integrated approach of machine-learning algorithms and hedonic modeling to inform equitable housing policies. *Land Use Policy*, *82*, 657–673. doi:10.1016/J.LANDUSEPOL.2018.12.030

Jackson, L. E. (2003). The relationship of urban design to human health and condition. *Landscape and Urban Planning*, *64*(4), 191–200. doi:10.1016/S0169-2046(02)00230-X

Jana, M. K. (2015). Visual pollution can have a deep degrading effect on urban and suburban community: A study in few places of Bengal, India, with special reference to unorganized billboards. *European Scientific Journal*, *7881*(June), 1–14.

Jana, M. K., & De, T. (2015). Visual pollution can have a deep degrading effect on urban and suburban community: A study in few places of Bengal, India, with special reference to unorganized billboards. *European Scientific Journal*, *7881*(June), 1–14.

Jellet, P. (1993). Beachwatch visual pollution ratings 1 July 1991 to 31 March 1993: Data analysis and modelling. NSW Environment Protection Authority.

Jensen, C. U., Panduro, T. E., & Lundhede, T. H. (2014). The vindication of Don Quixote: The impact of noise and visual pollution from wind turbines. *Land Economics*, *90*(4), 668–682. doi:10.3368/le.90.4.668

Jian, A. (2020). Road traffic safety managment of visual pollution by outdoor advertisements. eprints.utm.my. Retrieved from http://eprints.utm.my/id/eprint/85764/1/AngZhiJianMSKA2020.pdf

Johnson, M. G., & Symonides, J. (1998). *The universal declaration of human rights: A history of its creation and implementation, 1948–1998. In human rights in perspective*. Paris: UNESCO Publishing.

Jorge, A. P., & Gentil, P. A. B. (n.d.). Função social da propriedade: A problemática da poluição visual1 social function of property: The problem of visual pollution.

Jurate Kamicaityte-Virbasiene, O. S. (2013). Free standing billboards in a road landscape: Their visual impact and its regulation possibilities (Lithuanian case). *Eineerinvironmental Research Engineering and Management*, *4*(4), 66–78.

Kadir, J. J. (2017). Deformation and visual pollution in the commercial streets of the old city of Mosul. *Eurasian Journal of Science & Engineering*. Retrieved from https://core.ac.uk/download/pdf/227204443.pdf

Kamali, K., & Tahmouri, P. (2013). An analysis on urban beautification and its socio-economic effects. *World Applied Programming*, *3*(June), 232–235.

Kamičaitytė-Virbašienė, J., & Godienė, G. (2015). Analysis of legal and theoretical framework creating the methodology of visual pollution assessment for natural landscapes. *Journal of Sustainable Architecture and Civil Engineering*. Retrieved from https://www.ceeol.com/search/article-detail?id=483560

Kamičaitytė-Virbašienė, J., Godienė, G., & Kavoliūnas, G. (2016). Methodology of visual pollution assessment for natural landscapes. *Journal of Sustainable Architecture and Civil Engineering*, *13*(4). doi:10.5755/j01.sace.13.4.13820

Kan, R., Ngai, S., & Wong, C. Y. (2014). Visual pollution in Hong Kong. 2012.

Karimipour, H., Mojtahedi, M., & Dehkordi, F. A. (2015). Introduction to a quantitative method for assessment of visual impacts of Tehran Towers. *Journal of Soil Science and Environmental Management*, *6*(6), 132–139. doi:10.5897/JSSEM12.046

Kasenchak, R. T. (2019). What is semantic search? And why is it important? *Information Services & Use*, *39*, 205–213. doi:10.3233/ISU-190045

Kazakov, O. D., Kulagina, N. A., & Azarenko, N. Y. (2020). Machine learning methods in municipal formation. *Lecture Notes in Networks and Systems*, *73*, 339–346. doi:10.1007/978-3-030-15160-7_35

Kelly, M. (1998). *Encyclopedia of aesthetics*.

Khaled, M. T. J. (2009). Analyzing and evaluating visual pollution in Tulkarem city (center of Tulkarem city–case study). repository.najah.edu. Retrieved from https://repository.najah.edu/handle/20.500.11888/7484

Khanal, K. K. (2018). Visual pollution and eco-dystopia: A study of billboards and signs in Bharatpur metropolitan city. *Research Journal of English Language and Literature*. Retrieved from http://rjelal.com/6.1.18/202-208 KAMAL KRISHNA KHANAL.pdf

Kharate, S., & Banerjee, S. (2016). A study of visual pollution and its effect on mental health. *Innovation in IT*. Retrieved from http://www.indianjournals.com/ijor.aspx?target=ijor:jiit&volume=3&issue=1&article=008&type=pdf

Kieran, M. (2005). *Contemporary debates in aesthetics and philosophy of art*. Malden, MA: Blackwell.

Kietzmann, J. H. (2017). Crowdsourcing: A revised definition and introduction to new research. *Business Horizons*, *60*(2), 151–153. doi:10.1016/j.bushor.2016.10.001

Klein, N. (2000). *No logo*. Flamingo, Great Britian. Retrieved from http://dspace.vnbrims.org:13000/jspui/bitstream/123456789/1167/3/No Logo - Naomi Klein.pdf

Kim, J., & Kim, Y. (2019). Visual pollution on public sites affecting public sights. In The 32th KKHTCNN Symposium on Civil Engineering. Retrieved from https://koasas.kaist.ac.kr/handle/10203/276299

Kratz, J., & Littlejohn, G. (2007). Visual pollution. *Green Teacher*. Retrieved from http://search.proquest.com/openview/8b3bda9ffc55ef617035201aab10a998/1.pdf?pq-origsite=gscholar&cbl=33544

Kucharikova, Z., & Simko, J. (2017). Visual pollution localization through crowdsourcing and visual similarity clustering. In 2017 12th international workshop on semantic and social media adaptation and personalization (SMAP). Retrieved from https://ieeexplore.ieee.org/abstract/document/8022662/

Langer, S. K. (1953). *Feeling and form*. New York, NY: Scribner.

Lascu, L. F. (2015). The impact of visual pollution on quality of life. Case study: The historic core of Bucharest. *Calitatea*. Retrieved from http://search.proquest.com/openview/5831355d3b0b410d59c511413801b8cd/1?pq-origsite=gscholar&cbl=1046413

Loyola, D., & Ronel, B. (2018). *Berleant's non-cognitive engagement approach in remedying the problem on urban visual pollution*. Retrieved from https://www.dlsu.edu.ph/wp-content/uploads/pdf/conferences/arts-congress-proceedings/2018/acp-02.pdf

Lynch, K. (1960). *The image of the city*. The MIT Press. Retrieved from https://books.google.com.pk/books?id=_phRPWsSpAgC&printsec=frontcover&dq=effect+of+imageability+of+the+built+environment&hl=en&sa=X&ved=2ahUKEwibj8bu8rLtAhXJVBUIHTDYA0oQ6wEwAHoECAEQAQ#v=onepage&q&f=false

Maclurcan, D. C. B. (1973). Visual pollution. *Australian Surveyor*, *25*(2), 82–88. doi:10.1080/00050326.1973.10440658

Madleňák, R., & Hudák, M. (2016). The research of visual pollution of road infrastructure in Slovakia. In International Conference on transport systems telematics. Retrieved from https://link.springer.com/chapter/10.1007/978-3-319-49646-7_35

Marinescu, D., & Truşcă, P. (2010). Visual pollution and aesthetic education. *Metalurgia*. Retrieved from http://search.ebscohost.com/login.aspx?direct=true&profile=ehost&scope=site&authtype=crawler&

jrnl=04619579&asa=Y&AN=48618211&h=UDGsQ9it0BGCvBM%2FqnPyNZrMh%2BWBMrpJHd2dbgRreCHax0%2FJGxTa0%2BgsNsmcy%2B9UGybzhXCJAoN4KEm%2BBlhn2A%3D%3D&crl=c

Marks, P. (2009). Projector phones: Cool app or visual pollution. *New Scientist.* Retrieved from https://www.sciencedirect.com/science/article/pii/S0262407909605650

Mattila, H. (2016). Aesthetic justice and urban planning: Who ought to have the right to design cities? *Geojournal, 58*(2), 131–138.

McEvoy, J., III, & Williams, S. (1970). *Visual pollution in the lake Tahoe basin.* Davis, CA: University of California.

McEvoy, J., & Williams, S. (1976). *Visual pollution in the lake Tahoe basin: A report to the Tahoe regional planning agency based on the application of a quantitative method of assessing the visual environment.* Davis, CA: University of California.

McMahon, E. (1998). Litter on a stick. *Planning Commissioners Journal, 45,* 123.

Melhosseini, D. K., Mortazavi, S., & Hosseini, S. M. (2018). Investigating the visual pollution of non-heterogeneous land uses in the tourist areas (case study: Dohezar and Sehezar forests of Tonekabon). sid.ir. Retrieved from https://www.sid.ir/en/journal/ViewPaper.aspx?id=597316

Mendez, V., & Carmen, A. (2013). Visual pollution in public spaces in Venezuela. *Gestión y Ambiente.* Retrieved from https://inis.iaea.org/search/search.aspx?orig_q=RN:45026928

Mennan, Z. (2009). Basitten karmaşiğa: Gestalt algi kurami ilkelerinin karmaşiklik paradigmasi içinde sürdürülebilirliği. *Metu Journal of the Faculty of Architecture, 26*(2), 309–323. doi:10.4305/METU.JFA.2009.2.15

Meunier, S. (2010). *Visual pollution.* Göttingen: Steidl.

Minami, I. (2001). Historico sobre Publicidade na Paisagem – Paisagem Urbana de Sao Paulo, Publicidade Externa e Poluicao Visual (History of advertisement in the built environment – urban landscape of Sao Paulo, advertisement and visual pollution). Retrieved from http://www.vitruvius.com.br/arquitextos. Accessed on August, 2022.

Mishra, P. P. (2014, September). *Indus valley civilization. Encyclopedia of global religion.* doi:10.4135/9781412997898.n338

Moles, A. (1987). O Cartaz *(The poster).* Sao Paulo: Perspectiva.

Molina, J. R. (2006). *Public spaces or private places? outdoor advertising and the commercialisation of public space in Christchurch, New Zealand.* Christchurch: University of Canterbury.

Morano, P., Tajani, F., & Torre, C. M. (2015). Artificial intelligence in property valuations an application of artificial neural networks to housing appraisal. In *Advances in environmental science and energy planning* (pp. 23–29).

Morsink, J. (1999). *The universal declaration of human rights.* Philadelphia, PA: University of Pennsylvania Press. doi:10.9783/9780812200416

Mousa, I. (2011). Visual pollution in the commercial streets in Duhok city. High diploma thesis, Submitted to the Higher Institute of Urban Planning, the University of Duhok: Dahuk, Iraq.

Mowla, Q. A. (2011). Urban aesthetics: A study on Dhaka. *Asiatic Society Bangladesh Publication, III*, 169–186.

Muthukrishnan, N. (2015). Visual pollution-more dangerous than you think it is. C.P.R. Environmental Education System. Retrieved from http://cpreec.org/161.htm

Nadimpalli, M. (2017). Artificial intelligence risks and benefits. *International Journal of Innovative Research in Science, Engineering and Technology, 6*(6), 2–4. Retrieved from https://www.researchgate.net/publication/319321806

Nagle, J. C. (2009). Cell phone towers as visual pollution. *Notre Dame JL Ethics & Pub. Pol'y, 23*, 537.

Nagle, J. C., Adler, B., Appel, P., Barrett, A., Bellia, A. J., Bellia, T., … Vandenbergh, M. (2009). The idea of pollution. *Environments, 43*(November), 1–78.

Naguib, M. M. (2016). Visual pollution caused by banners and signage installed on buildings facades case study: Alexandria versus Moscow city. *International Journal of Scientific Engineering and Research, 7*, 1663–1667.

Nami, P., Jahanbakhsh, P., & Fathalipour, A. (2016). The role and heterogeneity of visual pollution on the quality of urban landscape using GIS; case study: Historical garden in city of maraqeh. *Earth and Environemntal Sciences, 6*(January), 20–29.

Nasar, J. L. (1988). In J. L. Nasar (Ed.), *Environmental aesthetics.* Cambridge University Press. doi:10.1017/CBO9780511571213

Nasar, J. L., & Hong, X. (1999). Visual preferences in urban signscapes. *Environment and Behavior, 31*(5), 671–691. doi:10.1177/00139169921972290

Nessim, A. A. (2020). Visual pollution: An approach to reduce the environmental impact of light pollution in Egypt, *Architecture and urbanism: A smart outlook*. Retrieved from https://link.springer.com/chapter/10.1007/978-3-030-52584-2_33

News, A. (2014, January 9). Al-Ahsa municipality, Aramco launch awareness campaign. *Arab News*. Retrieved from https://www.arabnews.com/node/506216/%7B%7B

Newsbeezer. (2018, April 23). Jeddah municipality continues visual pollution campaigns and raises 12,661 neglected vehicles. Retrieved from https://newsbeezer.com/ksaeng/jeddah-municipality-continues-visual-pollution-campaigns-and-raises-12661-neglected-vehicles/

NYC Global Parnters. (2011). *Best practice: Clean city Act report (issue April 2011)*. Retrieved from http://www.construction21.org/espana/articles/es/como-realizar-una-auditoria-energetica-y-no-perderse.html

OCE. (2014). *Annual report on the environment: Noise, light pollution and visual pollution*.

ODK. (2020). ODK collect. Retrieved from https://play.google.com/store/apps/details?id=org.odk.collect.android&hl=en&gl=US

Ogunbodede, E. F., & Sunmola, R. (2014). Posters, banners and billboards visual pollution in Nigerian urban environment: Challenges to urban managers. *IOSR Journal of Humanities and Social Science, 19*(6), 56–64.

Özyurt, M., & Kumbur, H. (2003). Visual pollution in Mersin and its environmental effects, 12th int. In Symp. on Environ. Poll. And Its Impact on Life in the Mediterr. Region, Antalya.

Parsons, J. A., Singh, G., Scott, A. N., Nisenbaum, R., Balasubramaniam, P., Jabbar, A., … Dunn, J. (2010). Standardized observation of neighbourhood disorder: Disorderoes it work in Canada? *International Journal of Health Geographics, 9*, 1–19. doi:10.1186/1476-072X-9-6

Phan, T. D. (2019). Housing price prediction using machine learning algorithms: The case of Melbourne city, Australia. In Proceedings – International conference on machine learning and data engineering, ICMLDE 2018 (pp. 8–13). doi:10.1109/ICMLDE.2018.00017

Planning, R., & State, E. (2020). Multivariate analysis of factors responsible for visual pollution in the central business district of ore town, Ondo State, Nigeria. *FUTY Journal of the Environment*, *14*(3), 20–34. Retrieved from https://www.ajol.info/index.php/fje/article/view/214770

Plummer, R. (2006). *Brazil's ad men face billboard ban*. BBC News. Retrieved from http://news.bbc.co.uk/go/pr/fr/-/1/hi/business/5355692.stm. Accessed on March, 2006.

Portella, A. A. A. (2007). *Evaluating commercial signs in historic streetscapes: The Effects of the control of advertising and signage on user's sense of environmental quality (issue July)*. (pp. 1–191). Oxford: Oxford Brookes University.

Portella, M. O. (2013). Environment and urban landscape: Advertising and outdoor advertising and visual pollution. *Revista Direito Ambiental e Sociedade*, *3*(2), 135–151.

Portella, A. A. (2014). *Visual pollution: Advertising, signage and environmental quality*. Routledge. doi:10.1080/02665433.2014.967496

Portella, A. (2016). *Visual pollution*. London: Routledge.

Portella, A. A., & Reeve, A. R. (2006, September). Visual pollution in historic city centres: Theoretical concepts to develop commercial signage controls in different cultural contexts. In *Books of abstracts of IAPS – international association for people-environment studies – environment, health and sustainable development* (p. 201). Alexandria.

Probst, A. K. (2016). Environmental visual pollution and its relationship to signage technology: A case study in Thailand. rave.ohiolink.edu. Retrieved from http://rave.ohiolink.edu/etdc/view?acc_num=ucin1459438017

Radomska, M., Yurkiv, M., & Nazarkov, T. (n.d.). The assessment of the visual pollution from industrial facilities in natural landscapes. Kdu.Edu.Ua. Retrieved from http://www.kdu.edu.ua/EKB_jurnal/2019_1(27)/PDF/45_49.pdf

Ramos, C. V., & Treviño, M. G. M. (n.d.). Perception of visual contamination of the population of the city of H. Matamoros, Tamaulipas. Mexico. Eumed.Net. Retrieved from http://www.eumed.net/rev/cccss/2017/03/contaminacion-visual-mexico.zip

Rescia, P., Corbetta, E., Puletti, M., & Rizzi, W. (2004). Offshore plants visual pollution mitigation. In SPE international conference on health, safety, and environment in oil and gas exploration and production. OnePetro. Retrieved from https://www.onepetro.org/conference-paper/SPE-86778-MS

Riyadh Environment. (2019). Eliminating the visual pollution and processing it. Retrieved from https://www.riyadhenv.gov.sa/en/plan/الحد-من-التلوث-البصري-ومعالجته/?tab=3,948

Robinette, G. (1972). Plants for easing visual pollution, or ways to overcome ugliness [beautification]. Yearbook of agriculture. US Department of Agriculture. Retrieved from http://agris.fao.org/agris-search/search.do?recordID=US19790421803

Roca, M. A. S. (2011). Visual pollution by television antennas on the Albaicin. A technologic proposal for its correction. *E-RPH-REVISTA ELECTRONICA DE PATRIMONIO HISTORICO*, *9*, 59–88.

Roca, M. A. S. (2016). Numerical assessment of visual pollution degree in an "cultural interest good": Methodology for implementation. Application of this method to the Albaicin neighbourhood of Granada. *E-RPH-REVISTA ELECTRONICA DE PATRIMONIO HISTORICO*, *18*, 70–83.

Rouse, A. C. (2010). A Preliminary taxonomy of crowdsourcing. In ACIS 2010 Proceedings – 21st Australasian Conference on Information Systems.

Russell, C. (2018). Sdg 11: Sustainable cities and communities from backyards to biolinks: Royal botanic. *BG Journal*, *15*(1), 31–33.

Sabah, O. A., & Samad, M. H. (2015). Livable heritage street and visual pollution in Georgetown/Penang. In Conference on liveable cities 2015 (ICLC2015). Retrieved from http://www.academia.edu/download/46329828/LIVABLE_HERITAGE_STREET_AND_VISUAL_POLLUTION_IN_GEORGETOWN-PENANG.pdf

Sachs, J., Kroll, C., Lafortune, G., Fuller, G., & Woelm, F. (2022). *Sustainable development report 2022*. Cambridge University Press. Retrieved from https://dashboards.sdgindex.org/chapters

Sahana, S., & Karthigayini, S. (n.d.). Design strategies to reduce the impact of visual and noise pollution in urban areas. *History*. Retrieved from https://core.ac.uk/download/pdf/327110692.pdf

Scenic America. (2000). *Scenic beauty benefits business: Design guidelines for business and historic districts*. Facts for Action. Washington, DC: Scenic America.

Schodorf, R. J. (1973). A study of visual pollution from overhead wires and associated structures. scholarworks.wmich.edu. Retrieved from https://scholarworks.wmich.edu/cgi/viewcontent.cgi?article=3701&context=masters_theses

Shaban, L. K., Suleiman, S., Abdel-Aziz, D., Isawi, H. Y. (2018). Evaluating the visual pollution in urban corridors-case of Al-Madina Al-Munawara corridor, Amman. *Research Journal of Applied Sciences, Engineering and Technology*. Retrieved from https://www.airitilibrary.com/Publication/alDetailedMesh?docid=20407467-201808-201811230017-201811230017-288-294

Shahzad, G. (2011). The impact of infrastructural services on traditional architecture and urban fabric of the Walled city of Lahore former director infrastructure sustainable development of Walled city Lahore, project. *Journal of Research in Architecture and Planning*, *10*(1), 35–44. Retrieved from https://jrap.neduet.edu.pk/arch-journal/JRAP_2011 (First Issue)/03.pdf

SilvaDa, M. (2020). Making sense of visual pollution: The "Clean City" law in São Paulo, Brazil. *Toxic truths*. Retrieved from https://library.oapen.org/bitstream/handle/20.500.12657/42642/9781526137005_fullhl.pdf?sequence=1#page=173

Silverstone, S. (1974). Visual pollution: Unaesthetic use of land as nuisance. *Alberta Law Review*. Retrieved from http://www.albertalawreview.com/index.php/ALR/article/view/2371

Sivaramanan, S. (2016). Visual pollution. *Sri Lanka Forester: Central Environmental Authority*, 26. Retrieved from https://books.google.com.pk/books/about/Visual_Pollution.html?id=bTkKuAEACAAJ&source=kp_book_description&redir_esc=y

Skhosana, M., Ezugwu, A. E., Rana, N., & Abdulhamid, S. M. (2020). An intelligent machine learning-based real-time public transport system. *Lecture Notes in Computer Science*, *12254*(LNCS), 649–665. doi:10.1007/978-3-030-58817-5_47

Somida, I. D. A. (2015). The effect of visual pollution on double meaning of the facad. fayoum.edu.eg. Retrieved from http://www.fayoum.edu.eg/english/engineering/architectural/pdf/THEEFFECTOFVISUAL.pdf

Souad, B. F. (2014). The impact of visual pollution on historic buildings (Tlemcen city, Algeria: Model). *Shedet*. Retrieved from https://shedet.journals.ekb.eg/article_87632.html

Stamps, A. (2000). *Psychology and the aesthetics of the built environment.* London: Springer.

Stapleton, R. M. (2008). Visual pollution. In 2008.

Stensland, S. M. (2004). Case note: Property law—Outdoor advertising control acts slice city funds into the bunker—In re denial of Eller media company's applications for outdoor advertising device permits in the city of mounds view. *William Mitchell Law Review, 31*(2), 10.

Stephany, T. J. (2010). *Energy Generation system for reduced visual pollution and cost.* US Patent App. 12/189,777. Retrieved from https://patents.google.com/patent/US20100032957A1/en

Sui, D., Elwood, S., & Goodchild, M. (2013). Crowdsourcing geographic Knowledge: Volunteered geographic information (VGI) in theory and practice. In *Crowdsourcing geographic knowledge: Volunteered geographic information (VGI) in theory and practice* (Vol. 9789400745, pp. 1–396). doi: 10.1007/978-94-007-4587-2

Sumartono, S. (2009). Visual pollution in the context of conflicting design requirements. *Journal of Visual Art and Design.* Retrieved from http://journals.itb.ac.id/index.php/jvad/article/view/742

Szczepańska, M., Wilkaniec, A., & Škamlová, L. (2019). Visual pollution in natural and landscape protected areas: Case studies from Poland and Slovakia. *Quaestiones Geographicae.* Retrieved from https://content.sciendo.com/view/journals/quageo/38/4/article-p133.xml

Taha, K., Obure, J., Al Ahmed, M., Al Dawsari, B., Ndugwa, R., Abilla, A., Oreido, W. K. D. (2019). The futue Saudi cities programme. CPI profile Jeddah. Retrieved from https://unhabitat.org/sites/default/files/2020/04/cpi_profile_for_jeddah_2019.pdf

Tama, B. (2020). Visual pollution in the urban environment. nda.rtu.lv. Retrieved from https://nda.rtu.lv/en/view/25589

Times, A. (2020, October 19). Hanging clothes in Balkonies banned. Retrieved from https://www.arabtimesonline.com/news/hanging-clothes-in-balconies-banned/

Tobergte, D. R., & Curtis, S. (2012). Urban advertising control in commercial streets: The case of Oscar Freire street. *III International Seminar NECC.* doi: 10.1017/CBO9781107415324.004

Tota, A. L. (2017). "Poluição visual": Uma reflexão sobre a sustentabilidade de imagens "visual pollution": A reflection on the sustainability of images. *RBSE REVISTA BRASILEIRA DE SOCIOLOGIA DA EMOÇAO*. Retrieved from http://www.cchla.ufpb.br/rbse/RBSEv.16n.47ago2017completo.pdf#page=105

Townsend, D. (1997). *An introduction to aesthetics*. Great Britain: Wiley Blackwell. Retrieved from https://www.amazon.com/Introduction-Aesthetics-Dabney-Townsend/dp/1557867313

Troy, L. (1974). Visual pollution in WA. Claremont Teachers College.

UN-Habitat. (2018). *Sustainable development goal 11. A guide to assist national and local governments to monitor and report on SDG goal 11+ indicators*.

United Nations. (2015). United nations sustainable development – 17 Goals to transform our world. In *United Nations*. New York, NY: United Nations Press.

Urban, M., Avilés, D. J. V., Bojović, M., & Urban, K. (2020). Artificial, cheap, fake: Free associations as a research method for outdoor billboard advertising and visual pollution. *Human Affairs*. Retrieved from https://www.degruyter.com/view/journals/humaff/30/2/article-p253.xml

Vaidya, H., & Chatterji, T. (2020). SDG 11 and the new urban agenda: Global sustainability frameworks for local action. In I. B. Franco, T. Chatterji, E. Derbyshire, & J. Tracey (Eds.), *Issue February* (pp. 173–185). Singapore: Springer. doi:10.1007/978-981-32-9927-6

Voronych, Y. (2013). Visual pollution of urban space in LVIV. *Space & Form*, 309–314.

Wakil, K. (2012). *A study of regulating Advertisement boards in Lahore* (pp. 1–70). Lahore: University of Engineering and Technology.

Wakil, K. (2022). Urban visual pollution: Assessment, practices, and management framework a case study of Rawalpindi, Pakistan. National University of Sciences and Technology, NUST, Islamabad, Pakistan.

Wakil, K., & Hussnain, M. Q. (2014). Regulatory mechanisms for outdoor advertisement boards in Lahore. In International conference on town planning and urban management (ICTPUM 2014) 1(1).

Wakil, K., Hussnain, M. Q., Naeem, M. A., & Tahir, A. (2016). Regulating outdoor advertisement boards; employing spatial decision support system to control urban visual pollution. In 8th IGRSM international conference and exhibition on geospatial and remote sensing.

Wakil, K., Hussnain, M. Q., Tahir, A., & Naeem, M. A. (2016). Regulating outdoor advertisement boards; Employing spatial decision support system to control urban visual pollution. *IOP Conference Series: Earth and Environmental Science*, *37*(1). doi:10.1088/1755-1315/37/1/012060

Wakil, K., Hussnain, M. Q., Waheed, A., & Naeem, A. M. (2016). Contextual review of outdoor advertisements: Impacts and regulatory practices. *Science International*, *1*, 531–535. Retrieved from http://www.sci-int.com/pdf/ 17789059681 a 64 531-535 Khydija Wakil, -+++REVIEWED -SS–NUST–ISD–30-1-15.pdf

Wakil, K., Naeem, M. A., Anjum, G. A., Waheed, A., Thaheem, M. J., & Hussnain, M. Q. (2019). The assessment and mapping of urban visual pollution through an assembly of open source geospatial tools. *Real Corp 2019 Proceedings*, *4*(April), 723–730.

Wakil, K., Naeem, M. A., Anjum, G. A., Waheed, A., Thaheem, M. J., Hussnain, M. Q., & Nawaz, R. (2019). A hybrid tool for visual pollution assessment in urban environments. *Sustainability*, *11*(8), 2211. doi:10.3390/ su11082211

Wakil, K., Tahir, A., Hussnain, M. Q., Waheed, A., & Nawaz, R. (2021). Mitigating urban visual pollution through a multistakeholder spatial decision support system to optimize locational potential of billboards. *ISPRS International Journal of Geo-Information*, *10*(2). doi:10.3390/ijgi10020060

Wang, J. (2018). Spatial sampling. In *International encyclopedia of geography* (pp. 1–6). Wiley. doi:10.1002/9781118786352.wbieg0524.pub2

Wang, J.-F., Stein, A., Gao, B.-B., & Ge, Y. (2012). A review of spatial sampling. *Spatial Statistics*, *2*, 1–14.

Wilts, H., Garcia, B. R., Garlito, R. G., Gómez, L. S., & Prieto, E. G. (2021). Artificial intelligence in the sorting of municipalwaste as an enabler of the circular economy. *Resources*, *10*(4), 1–9. doi:10.3390/resources10040028

Xiangzhan, C. (2008). Urban image and urban aesthetics: Urban aesthetics in cross-cultural perspective. *Urban Image and Urban Aesthetics: Urban Aesthetics in Cross-Cultural Perspective*, *25*(2), 59–71.

Xiaolan, T., Yanyan, J., & Wenyuan, B. (2018). An analysis of visual pollution Problems and countermeasures in the construction of beautiful urban-rural areas. *Journal of Nanjing Forestry University*. Retrieved from http://en.cnki.com.cn/Article_en/CJFDTotal-NJLS201802008.htm

Xin, S. U. N. (2011). Discussion of visual pollution in city. *China Environmental Management*. Retrieved from http://en.cnki.com.cn/Article_en/CJFDTotal-HJGN201104019.htm

Xiu, S., Zhou, Q., Cai, Q., Long, L., Lu, D., Xiang, S. (2009). Investigation and research of university campus visual pollution. *Bulletin of Science, Technology & Society*. Retrieved from http://en.cnki.com.cn/Article_en/CJFDTotal-KJTB200903024.htm

Yao-zhi, L. (2011). The analyses of visual pollution problem and its strategies in historic district. *Modern Urban Research*. Retrieved from http://en.cnki.com.cn/Article_en/CJFDTotal-XDCS201103015.htm

Yellin, I. E. (1971, September). Visual pollution and aesthetic regulation. *Los Angeles Bar Bulletin, XLVI*, 11.

Yilmaz, D., & Sagsöz, A. (2011). In the context of visual pollution: Effects to Trabzon city center Silhoutte. *Asian Social Science*, 7(5), 98.

Zaeimdar, M., Sarab, F. K., Rafati, M. (2019). Investigation of the relation between visual pollution and citizenry health in the city of Tehran (case study: Municipality districts No. 1 & 12 of Tehran). *Anthropogenic Pollution*. Retrieved from http://ap.iauardabil.ac.ir/article_546040.html

Zaydee, N. M. Al. (2013). Visual pollution in the city of Mosul (A study in geographic contamination). *Mosoliya Studies*. Retrieved from https://www.iasj.net/iasj?func=article&aId=76673

Zbadi, H. (1997). *An analytical study of the symptoms and causes of visual pollution in the contemporary Egyptian city*. MA thesis, Department of Architecture.

Zhou, D., Wang, B. J., & Shi, B. (2011). GIS viewshed analysis of visual pollution assessment for mine environment. *Guilin Gongxueyuan Xuebao/Journal of Guilin University of Technology*. Retrieved from http://en.cnki.com.cn/Article_en/CJFDTotal-GLGX201102009.htm

INDEX